WHATEVER THE WEATHER

Science Experiments and Art Activities That Explore the Wonders of Weather

ANNIE RIECHMANN AND DAWN SUZETTE SMITH

BOULDER
2016

Roost Books
An imprint of Shambhala Publications, Inc.
4720 Walnut Street
Boulder, Colorado 80301
roostbooks.com

9 8 7 6 5 4 3 2 1

First Edition
Printed in The United States of America

♾This edition is printed on acid-free paper that meets the American National Standards Institute Z39.48 Standard.
♻Shambhala Publications makes every effort to print on recycled paper. For more information please visit www.shambhala.com.

Distributed in the United States by Penguin Random House LLC and in Canada by Random House of Canada Ltd

Designed by Danielle Deschenes

Library of Congress Cataloging-in-Publication Data

Riechmann, Annie, author.
Whatever the weather: science experiments and art activities that explore the wonders of weather/Annie Riechmann and Dawn Suzette Smith.—First edition.
pages cm
Includes bibliographical references.
ISBN 978-1-61180-231-3 (pbk.: alk. paper)
1. Weather—Juvenile literature. 2. Weather—Experiments—Juvenile literature. 3. Handicraft—Juvenile literature. I. Smith, Dawn Suzette, author. II. Title.
QC981.3.R54 2016
551.6078—dc23
2014048182

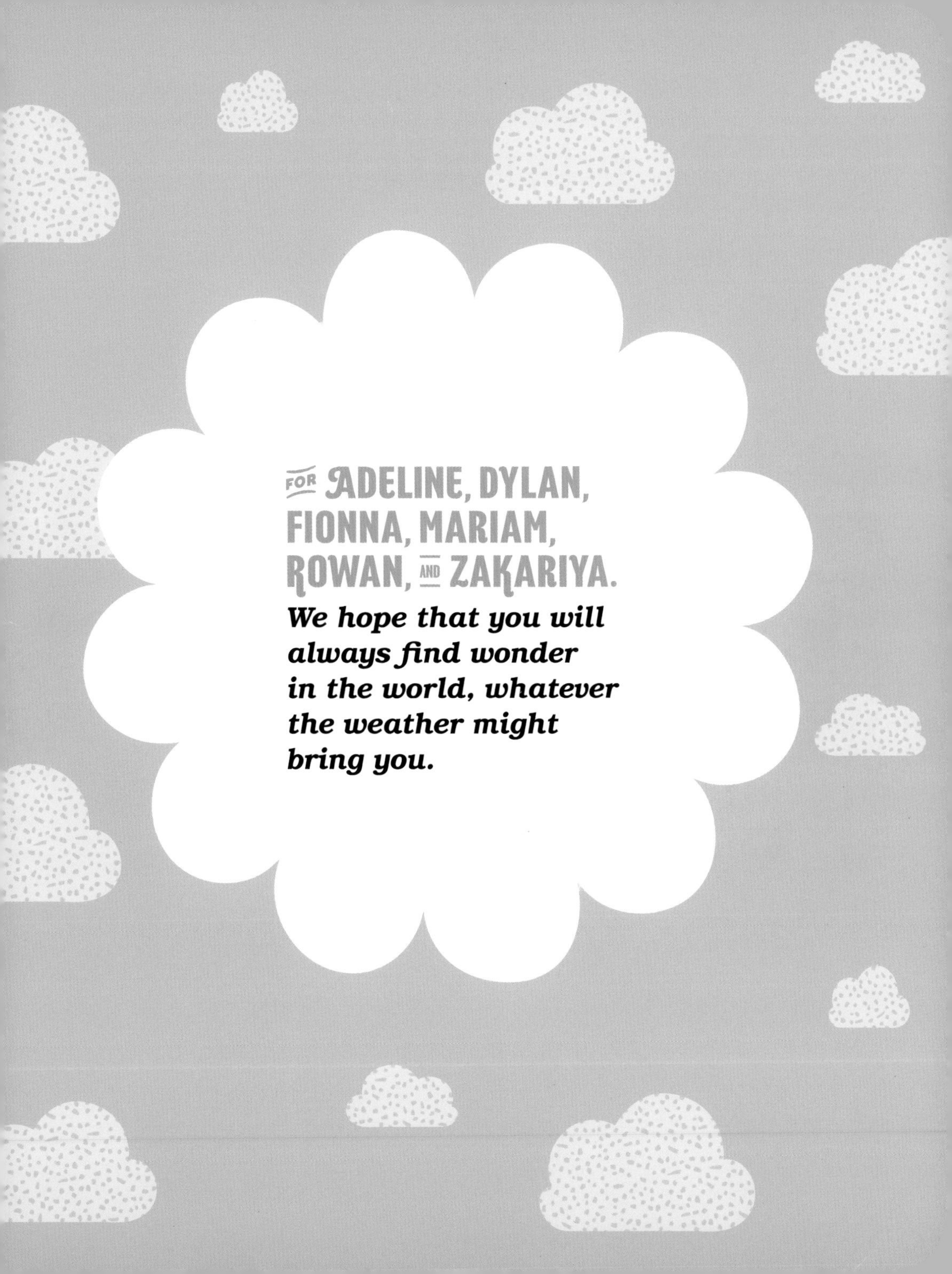

FOR ADELINE, DYLAN, FIONNA, MARIAM, ROWAN, AND ZAKARIYA.

We hope that you will always find wonder in the world, whatever the weather might bring you.

CONTENTS

Introduction 1

A Parent's Guide to Weather Exploration 5

Chapter 1. Baby, It's Cold Outside
15

Chapter 2. Saving for a Rainy Day
67

Chapter 3. Let the Sun Shine In
115

Chapter 4. Where the Wind Blows
157

Acknowledgments	198
Resources and Further Exploration	200
About the Authors	207

INTRODUCTION

Weather is a bit of a strange thing. ***It is such a part of our everyday routines that we often go about our business without even considering it in a truly conscious way. But at the same time, our daily lives are built with the weather at their heart. A beachside picnic during a January blizzard is unlikely to find a place on the calendar when each winter rolls around, for example. We change our tires, our wardrobes, our party plans, all because of what the weather brings each day and each season. But how often do we really explore the weather and what makes it shift, surprise us, and set the direction of what we do and when we do it? Basic discussions of the weather are common in the average elementary classroom, but often those discussions start and stop at the "what" of the weather, leaving out the much more interesting "why."***

The pages of this book aim to take a deeper look at the "why" of the weather, while also encouraging families to use their imaginations and their sense of wonder about the world to guide them through interesting open-ended art projects and some very entertaining science. The projects and experiments in this book are simple in conception, with the hope that they will be accessible and enjoyable for even the busiest families, and that they can be easily adapted to the creative sensibilities of each individual. More than anything, we'd like to see hands-on science and artistic endeavors become a casual part of your active family lives. And if you and yours learn some fascinating facts about the way the weather in our world works while you are tinkering, creating, observing, and wondering? Well, that is all the better.

The world that we live in is a collection of shadows and storms, wind and water. Each day brings sun, clouds, rain, or breezes our way, and regardless of temperature or wind speed, the opportunities for exploration are never limited. ***Whatever the Weather*** doesn't aim to solve all of the world's meteorological mysteries in one go. But it will help bring the basics of weather science down to earth for curious kids, taking them from thinking only about what weather does for us into thinking about what they can do ***with*** the weather. Each chapter of the book is organized around a common element of the weather, and the activities and projects inside focus on taking that element and transforming it into a medium for art, observation, and experimentation. Paint with raindrops, watch the sun make pictures, explore evaporation while making an evolving work of art. Experience the power of air pressure with the quirkiest of kitchen-table science projects, or use the incredible power of intense cold to create unique building materials that will please emerging architects to no end. This book provides ample evidence for the idea that the weather allows us to do almost anything if we think creatively and approach each day outdoors with a sense of adventure and a hankering for exploration.

Whatever the Weather is certainly a mixed bag of a book. There is art, there is science, there is exploration just for the sake of satisfying undeniable childhood curiosity about the world and the way that it works. This book operates on the assumption that with the proper inspiration, each of us is an explorer, an artist, and a scientist. There is something for each member of your family in these pages. We hope that this book will help your family to see that even an ordinary day is full of mystery, magic, and plenty of opportunities to make the most of whatever the weather may bring.

Why We Wrote This Book

We came to the writing of this book in a meandering sort of way. For a few years, the two of us worked together on a variety of diverse projects based on the ever-changing interests of ourselves and our growing children. We first began writing together on a collaborative blog along with a handful of other parents who were also focused on sharing the natural world with their children through simple, authentic experiences. Life got busy for everyone involved. Things like new babies and cross-country moves made for big changes, and that original project eventually wound down. But the two of us had found each other and made fast friends in the process.

From there, we shared the work of producing occasional special posts for one another's personal online spaces, and eventually began a new blogging partnership together at our current Internet home, ***Mud Puddles to Meteors***. Despite living hundreds of miles apart, we developed a strong kinship through emails, phone calls, exchanging snail mail, and a shared dedication to finding ways to fold natural history and simple science into our own family lives and those of our readers. We both experienced a number of significant shifts in the landscapes of our personal and family lives, and through it all, helped each other to remember how well anchored we were in our appreciation for the reliable cycles of the natural world.

At some point along the path of our writing and working together, we began to realize that one of the defining features of the outdoor excursions we took with our kids, whether it was for purposes of family fun or for more formal learning, was that the success (or lack thereof) of these adventures was ultimately tied to the weather. It was no good to get outside with a freshly formed baking soda clay volcano on a rainy day, nor did it make sense to go bug hunting with clumsy, mitten-clad fingers in the snow. We realized that the cooperation of the weather and our ability to do the planned activity in it seemed not to matter at all to our collective crew of kids. Indeed, they were as interested in what they could do with the weather itself as they were in the projects we had originally planned, and the simpler the investigation of current conditions, the better. They often took the materials we had planned to use for a particular project and used them to create in new and unexpected ways. We watched as our kids played in the rain, built with the snow, and chased their own shadows in the sunshine. We observed them as they made discoveries about wind direction, the size of raindrops, and how snow melts, without even knowing that was what they were doing. And we

realized that for kids, without jobs or obligatory errands to run, the weather is a big part of what structures life. We began exploring weather science alongside our children in earnest, following their lead, and found that we loved the experiences as much as they did.

Through a series of very fortunate events, we soon found ourselves fully immersed in the writing of this book: a manual to experiencing the wonder of the weather with a curious heart and open hands. Although it might not be the book we expected to pen when we had our first optimistic phone conversation about undertaking this kind of endeavor together, in time, it became the one that made perfect sense.

About the Activities

Before you begin thumbing pages and preparing to dive in, it seems helpful to share a few thoughts about the activities here in the book. First of all, it is important to note that the activities in these pages are designed to be simple in both conception and execution. We hope that you will find that they allow for spontaneous science and art to become a part of your family life. Think of the activities in this book as inspiration, not a prescription. Use our ideas and instructions as a place to start, knowing that we absolutely intend for curiosity and a spirit of exploration to lead you off the path of the activities from time to time. In fact, we hope that is what happens!

As you flip through the book, you'll notice that there are a couple of features that appear regularly on the pages. One of these, the "Science behind the Scenes," accompanies each of the activities. Think of these boxes as a parent's cheat sheet to the weather science related to the activity at hand. You will find fascinating facts and interesting information that will help you explain things to your children when questions arise. And you might very well learn something new, too! Scattered throughout the book are also "Wondering about Our World" features. These are not so much full-fledged projects as they are inspiration for families who find themselves inclined to dig deeper and explore a bit more.

We hope that you enjoy this book just as much as we loved writing it!

A PARENT'S GUIDE TO WEATHER EXPLORATION

As you look forward to beginning your family foray into weather exploration, we thought that it might be helpful to provide a few pointers, as well as a sense of the supplies that might be good fun to have along on the journey. In the pages that follow, you'll find instructions for putting together a supply kit that you and yours can keep at the ready, ensuring that your weather exploration outings are a rousing success each time. We've also provided an overview of some of the simple weather science principles behind the project and activities in the book. Think of it as a grown-up's guide to weather science basics and an excellent way to be prepared for questions that your young people might pose as you move through the pages of the book together.

Whatever-the-Weather Supply Kit

The projects found in this book are generally straightforward and require minimal supplies. While the necessities are things you can typically gather from your art supply box or kitchen junk drawer, we have found that having many of the materials already corralled in a watertight bag or bin has encouraged us to take them with us to have available at the beach, park, or even just for heading out to the yard. Here is a handy list of items used in the activities and experiments in this book. You can use this as an easy checklist to assemble a quickly accessible weather exploration supply kit of your own. This way, you can help your kids wrestle with their rain gear or snow boots, knowing that your supplies are at the ready and it will otherwise be easy to get out the door to conduct experiments, make art, and have some fun at a moment's notice.

SIMPLE SUPPLIES FOR WEATHER EXPLORATION

- TUBE WATERCOLOR PAINTS IN PRIMARY COLORS
- PAINTBRUSH
- GLUE
- NEEDLE
- THREAD
- WATERCOLOR PAPER
- ORIGAMI PAPER
- PUSHPINS
- DUCT TAPE
- SMALL PIECE OF DARK FABRIC
- YARN
- STRING

A Note about Quality Supplies for Young Artists

We encourage you to invest in the highest quality art supplies you can afford for your children, when possible. Using high-quality supplies can make a big difference in the outcome of a project. They tend to be smoother in application and easier to use (think of the frustration of using cheap packing tape that keeps ripping and getting stuck together), and this alone can encourage children to explore more with the supplies at hand. There is also something to be said for demonstrating faith in your child's artistic abilities through allowing them to use "real" materials. A few lessons about respecting supplies and coaching about how to use or apply various mediums will be needed to help them learn proper usage, but it will pay off in the end with a more engaging experience and pleasant results.

Weather Pen Pals

In some ways, working your way through the projects and activities in this book is a little like taking a journey with your child. Along the way, you will learn new things and try your hand at creating interesting artifacts from the experiences you have together. We thought that perhaps you'd like to invite a friend along.

Have you ever talked to a friend or relative and found out that while it was raining at your house they were looking up at blue sky and sunshine? Weather is obviously different in one place than in another, and talking about the weather with people living in other locations can be a great way to get perspective on what life is like other places. See if your child would like to ask a friend or relative to start taking notes about the weather where they live. Then, decide on a time to compare notes. This can be once a month or even once a week.

Try to do this for a whole year to compare the similarities and differences. See if you can discover the changing weather patterns for your regions and why they are the same or different. You might also think about sharing some of the resulting artwork that comes from working your way through the pages of this book. You can exchange snowflake photos, trade prints made from rainy-day leaf printing, or share data gathered from individual rain gauges to set outside your home.

THINGS TO WRITE ABOUT WITH YOUR WEATHER PEN PAL

- WHO GETS MORE RAIN?
- DOES THEIR REGION HAVE A DIFFERENT "RAINY SEASON" THAN YOUR LOCATION?
- WHAT IS THE TEMPERATURE WHERE YOU LIVE? THE HUMIDITY? HAS THERE BEEN SNOW?
- HAVE THERE BEEN BIG STORMS OR HIGH WINDS?
- DID YOU HAVE PLANS THAT WERE CHANGED BECAUSE OF THE WEATHER?
- WHAT KIND OF WEATHER HAVE YOU BEEN WISHING FOR?
- WHAT KIND OF CLOUDS HAVE YOU BEEN SEEING?
- HOW DOES THE WEATHER AFFECT PLANTS AND ANIMALS NEAR YOUR HOME?

The Water Cycle

As you move through the projects in this book you may find that your child has questions that extend beyond the more basic observations made during the activities and experiments. And since this book is about the weather, many of the projects are related to one of the magical forces that makes our

world work: the water cycle. While we provide plenty of weather science to accompany the projects and activities in the book as you do them, helping your child understand the basics of the water cycle will go a long way toward encouraging them to ask pertinent questions and to use their background knowledge to arrive at their own answers. And since we adults can always use a refresher when it comes to things we first learned as kids ourselves, we hope that this mini–crash course on where water comes from and where it is going will help you, as a parent, caregiver, or teacher, to be able to point out relevant phenomena in the moment as a way to inspire inquiry and wonder. To that end, we have provided a primer here to help break down the water cycle and provide a basic outline of how it works.

In many ways, weather is water. Rain, snow, and clouds hovering in the sky overhead are all parts of our weather patterns, and they are all the result of a very important global system called the water cycle. Although much of the science presented in this book is of the more hands-on and experiential variety, we thought that it made good sense to explore the water cycle before we fully dive in. The water cycle, after all, is what makes so much of the weather-based fun we are about to have possible in the first place.

The water cycle, or hydrologic cycle, demonstrates the constant movement of water on Earth. Through the cycle water changes state, or form, from liquid, to vapor, to solid and back again, over and over.

There are four main processes in the water cycle:

EVAPORATION, SUBLIMATION, OR TRANSPIRATION

CONDENSATION

PRECIPITATION

COLLECTION

For the water cycle to begin water must enter the air. But how does water get into the air all around us? There are a few different ways:

Most of the water in the air is transported there through the process of evaporation. This is the process by which liquid water turns into water vapor, or steam. It begins when liquid water heats up and the water molecules, the tiny units that make up water, begin to bump into each other. Some of these molecules are knocked out of the liquid and enter the air above. They become water vapor. When you see steam rise from a cup of hot chocolate on a cool winter day, you are watching the process of evaporation. Most of the water vapor in the air comes from the oceans, lakes, and rivers of the world.

Water vapor can also enter the air through a process called ***sublimation***. Sublimation occurs when water turns from a solid (like ice) to a vapor. Have you ever seen a patch of ice in your backyard get smaller even though the temperature was still below freezing? This shrinking ice was not melting. Instead, it was losing water to the air. That is sublimation at work!

But evaporation and sublimation are not the only ways that water vapor gets into the air. Water can also enter the air through transpiration. ***Transpiration*** is the process by which water enters the air from the leaves of plants.

If you have ever left a glass of cold water sitting outside on a hot summer day, you might have noticed that drops of water begin to appear on the outside of the glass. Your child might wonder: Was the glass leaking? Where

was the water coming from? These are good questions! Actually, the water was coming from the air all around it. Water vapor in the air was "condensing" on the outside of the glass, because when water vapor gets cold it changes back into a liquid. When the water vapor in the warm summer air came into contact with the cold glass, it became liquid again. In nature, clouds are formed this way. ***Condensation*** is an important part of the water cycle because it is responsible for turning water vapor into liquid water again. When the water vapor becomes a liquid, it can once again make its way back to the surface of the Earth.

Most water is returned to the surface of the Earth in the form of rain, but it can also return as snow, sleet, hail, or even freezing rain. When water falls from a cloud to the surface of the Earth it is called ***precipitation***.

Water that falls back to the Earth as precipitation is then collected to begin the cycle again. In the process of ***collection***, some of this water will simply happen to reach the surface of the Earth in areas already covered by water, such as rivers, lakes, or the vast areas of ocean. Precipitation water that reaches the dry land will be absorbed by the Earth, or may run off into local bodies of water.

Now that you have assembled your tool kit and revisited some of the science concepts that you may not have thought about in a while, you should be all set to start experimenting, creating, and observing. The projects in the pages that follow will encourage you to do just that, with an eye on the sky, and your face to the wind. The weather on any given day may not be ideal for picnicking, baseball in the park, or going for a swim. But we hope that using this book as a guide, you and yours will begin to see that science, art, and creativity can be on the day's agenda no matter what the weather report has to say.

The Beaufort Scale

Think of the wind as one of the main characters in this book of ours. It does, after all, have its own chapter! To this end, it makes good sense to look a bit at how the wind is measured, particularly because the origin of our modern meteorological measurement system is perhaps unexpectedly quirky and fun. In other words, let's talk about the Beaufort scale.

The Beaufort scale was created by Sir Francis Beaufort in the early 1800s as a way to standardize the observation of wind speed made by sailors during voyages at sea. It was later adopted by the Royal Navy. While the World Meteorological Organization now relies on more modern equipment to help with weather observations, variations of the scale are used by most countries around the world when reporting weather conditions. The scale started out giving ratings to the effect the wind had on the sails of a ship at sea, but over the years it has changed to include wind speed, sea conditions, and visible impact on land. While the scale continues to evolve, ratings from 0 (calm) to 12 (hurricane) help distinguish a gentle breeze (force 3) from a fresh gale (force 8). Along with descriptions such as "light air," "strong breeze," and "storm," the scale also gives an idea of what might be happening around the observer to help classify the wind force. For instance, a force 0 would be so calm one could see reflections in a body of water; during a force 2 leaves begin to rustle; branches of trees sway in a force 5; and waves crash on shore causing damage during a force 11 wind.

The following table can help you describe wind force and the effects of wind on the surrounding environment to your child. Use the table to make predictions of wind speed and check your local weather to see if your calculations were accurate. Remember that gusts and local land features may alter the wind patterns in your specific location so your wind speed may vary from the location where the local meteorologist gathered information for the regional forecast. If this is the case, it would be good to talk about what types of things would impact your immediate location, such as a large hill behind your house, being next to the sea, or being at a slightly higher elevation.

FORCE	DESCRIPTION	WHAT'S HAPPENING	WIND SPEED
0	Calm	See reflections on water, smoke rises up	0 mph
1	Light air	Ripples appear on water, smoke drifts in the direction of the wind	1–3 mph
2	Light breeze	Small wavelets form but do not break, leaves rustle	4–7 mph
3	Gentle breeze	Large wavelets form crests with scattered whitecaps, twigs on trees move	8–12 mph
4	Moderate breeze	Small waves form with frequent whitecaps, leaves blow	13–18 mph
5	Fresh breeze	Moderate waves form with many whitecaps and chance of spray, branches on trees sway	19–24 mph
6	Strong breeze	Large waves form crests with some spray, larger branches on trees move	25–31 mph
7	Near gale	Sea heaps up with white foam from breaking waves blown in streaks in direction of wind and moderate spray, larger trees sway	32–38 mph

FORCE	DESCRIPTION	WHAT'S HAPPENING	WIND SPEED
8	Fresh gale	Moderately high waves with breaking crests forming spindrift with considerable spray, twigs break off trees	39–46 mph
9	Strong gale	High waves with crests that sometimes roll over, spray reduces visibility, shingles blow off rooftops	47–54 mph
10	Whole gale	Very high waves with overhanging crests, foam makes sea take on white appearance, spray reduces visibility, trees blow over	55–63 mph
11	Storm	Exceptionally high waves, large patches of foam cover sea surface, spray reduces visibility, waves crash on shore causing widespread damage	64–72 mph
12	Hurricane	Huge waves, sea almost completely white with foam, visibility greatly reduced, general widespread destruction on land	72+ mph

Now that you are familiar with the Beaufort scale, you might want to take some time to look around your own yard or neighborhood to find trees and other markers to keep an eye on when the wind whips up. Use these to decide what the wind speed might be at the time. You can even use these to make your own wind scale that you can write up and tuck into your supply kit to have at the ready for a windy day, or anytime that you want to record the weather conditions.

1

BABY, IT'S COLD OUTSIDE

Of the four seasons, winter is the one perhaps most often associated with an air of enchantment and anticipation. It is during the winter months that the familiar landscapes of our everyday lives become transformed by cold, often in surprising ways. Snow may cover parks and yards, muffling normal neighborhood noises and making the world seem quiet and still. Or perhaps you'll wake to find ice blanketing the branches of your favorite backyard tree, causing it to sparkle in the sunshine even though the temperatures are freezing outside. These dramatic changes and the ways in which they make the world appear new and different are one reason why winter is a great time to get outside and experience the wonders of ice and snow firsthand.

Not to take too much away from the magic of the season, but did you know that all of that wonderful sparkle, soft snow, and frosty ice is simply the result of water and the way that it behaves in cold weather? Indeed, as we know from our brief introduction to the water cycle at the beginning of this book, water changes state when there are changes in temperature. When water cools to near freezing temperatures, it becomes solid in the form of snow or ice. So how is snow different from ice? Why is thin ice clear or "see-through" while snow is white? Why do icebergs and glaciers sometimes appear blue? The activities in this chapter will be your guide as you explore more about the way that cold weather changes the water in the world around us and brings the idea of a winter wonderland to life.

The projects and activities in this chapter have been created with the joyful exploration of nature's winter bounty in mind. These projects use materials provided by the natural environment to experiment with weather science and the remarkable changes cold makes in our environment. Gather ice and snow to build simple structures that sparkle in the sunlight. Make giant frozen "marbles" and use them to create towers, walls, castles, and more. Photograph snowflakes and use them as a makeshift thermometer. Pull on a pair of mittens and get creative. Winter is waiting.

ICE SHEET SCULPTURES

The thin sheets of ice that form across the tops of puddles and other small bodies of water are one of the more unusual building materials that nature provides. It might not be the first thing to come to mind when you see a driveway puddle that has frozen over, but the thin sections of ice that can be pulled away from that puddle can be used to make wonderfully unique ice sculptures. A few pieces of flat ice and a bit of water are all that you need to begin work on a beautiful piece of abstract art that will last as long as the freezing temperatures (or small stomping feet) allow.

WHAT YOU WILL NEED

- ICE SHEETS FROM A PUDDLE, POND, OR STREAM
- CONTAINER (OPTIONAL)
- WATER

SCIENCE BEHIND THE SCENES: *Origins of Ice*

As you and your child begin to explore the remarkable nature of ice together, you might find that your child has questions about what ice is and why some ice is different in shape, appearance, or texture than other types of ice you encounter. Simply put, ice is water frozen solid. When water is cooled to temperatures below 32 degrees Fahrenheit (or 0 degrees Celsius), it freezes and becomes solid. The color of ice, as well as how see-through or transparent it is, depends on what was in the water that made the ice before it became solid and how dense it is. If there were minerals or impurities in the water, those will be found in the ice as well. Ice can come in many forms, from the cubes in your freezer to hail, frost, snowflakes, and even giant formations such as the polar ice caps. Ice is a very important part of the water cycle.

WHAT YOU WILL DO

Begin by gathering sheets of ice, choosing a larger one to use as the base for your work to sit on. If you'll be setting up for building in an area far from your source of ice, you might think about collecting the ice pieces in a container that is easy to tote back and forth as needed. Summer beach buckets tucked away in the garage or under the back stairs might come in handy during the winter, as kids love to haul buckets of ice.

Be sure to use extra caution when gathering ice from the banks of larger bodies of water such as ponds, lakes, streams, and rivers. There is always a danger of falling into the frigid water, so in this case, the ice-gathering portion of the project might be best suited for the adult in the crowd, leaving the kids to be the bucket brigade transporting ice to a safe building location.

To begin, dip one edge of a smaller ice sheet into the water and place it on the base. Hold it steady until it freezes into place. This can be an exercise in patience for younger builders, but as long as the outdoor temperature is low it should not take too long.

To build your sculpture upward, drip water on the section you are building onto and then gently place the additional ice

sheet on top. It will freeze into place after a few moments time. How long this takes will depend on the temperature outside, so again, be patient and hum a tune while you wait if need be.

Continue using this method of placing ice sheets until you are satisfied with your sculpture and feel that it is complete. At this point, you might think of snapping a quick photo or of finding a warm, dry place where your child can make a sketch of their architectural masterpiece. Weather changes quickly, and ice often comes and goes just as fast!

If you have access to nearby snow, you can use it as a foundation to make your sculpture building go more smoothly. Simply anchor the initial pieces of ice in the snow, pushing them in as deep as you need to for them to stay upright on their own.

In nature, ice often forms with dirt or other impurities trapped inside. Making your own ice sheets and mixing in a variety of materials is an interesting way to see how this process works in nature while adding a bit of visual flair to your ice. Try pouring water onto a cookie sheet, and before sliding the sheet into the freezer or placing it outside in freezing temperatures add dirt, food coloring, shredded leaves, seeds, or winter berries.

If you have ice sheets leftover after building, try using them as a lens to look at different objects around the yard. Are you able to see through the ice? Does looking through the ice change the way things on the other side look? Does the thickness of the ice seem to matter? Or the texture? Try using different pieces of ice to see how different the resulting view can be.

If you find that most of your ice is uniform in texture or thickness but you'd like to mix it up a bit, try letting the ice melt a little before setting it out to refreeze. Or break the ice up into small pieces, mix those with water, and then freeze again before building. You can make thicker pieces of ice by letting pieces melt slightly (or by pouring water over them) and then stacking another piece on top before refreezing. This would be a great way to make a tower, too!

WINTER FAIRY CASTLES

This project has an undeniable element of magic to it. Sparkling icicles and bits of nature found close by are used to construct fairy castles that shimmer in the winter sunlight and that can be customized according to the imagination of the maker. If you are in a cold climate, you can rely on Mother Nature to provide you with a set of icicles for building with. If they are not currently growing on the roofline of your home, a running creek with small waterfalls is a great place to find icicles "in the wild" (they grow just about anywhere you can find some trickling water and cold temperatures). Take the opportunity to go on an icicle hunt and encourage your child to look at where and how icicles are formed. Are they smooth? Do they have ripples? Are they straight or crooked? What would cause them to form differently? You do not have to have all of the answers to these questions. The simple act of inquiry is worthwhile in and of itself, and may inspire your child to do some investigation of their own. Of course, where there are icicles there is ice, so the key is to be careful when collecting. Falling icicles are a hazard and therefore should only be collected when it is safe to do so.

There are also instructions here for making a set of icicles yourself in the event that the weather doesn't quite cooperate with plans for construction, or if you find that the icicles around your home are too difficult to reach and gather.

WHAT YOU WILL NEED

- ICICLES, EITHER FROM THE WILD OR MADE AT HOME ***(see instructions on page 27)***
- ITEMS FROM NATURE SUCH AS CONES, DRIED SEED HEADS, AND BARK

SCIENCE BEHIND THE SCENES: ***Growing Icicles***

When looking at icicles, it is often not obvious to the casual observer just how they are formed. It is, after all, a bit of a slow process. Once they have moved beyond being fascinated by the beautiful appearance of icicles, kids often become curious about the natural processes that bring them into being. Icicles are a common ice formation made when dripping water freezes before it can reach the ground, and they can form anywhere there is dripping water and a cold enough temperature. As each drop of water trickles down and freezes it adds another layer of ice to the icicle, making it grow larger. It is possible for icicles to get so large that they grow all the way from a high roof down to the ground below, and some icicles can also be sharp and even dangerous if they were to fall.

WHAT YOU WILL DO

Help your child carefully transport the icicles that have been gathered to a good building location. Some icicles may break during the gathering process, or on the trip to the construction site, but they can still be used.

Start by creating the outline of your castle. If you have snow around, it is easy to stick the icicles in the snow for your base. It may help to use a small stick to start the holes, so as not to break the icicles when punching them through a layer of crusty snow. If there is no snow but the temps are cold, find a cold, flat surface such as a rock or sidewalk puddle to build on. Wet the bottom of the icicles with a bit of water and hold them in place until they freeze upright. As with other ice building endeavors, a little patience is required.

Once the basic structure of your castle is in place, gather up bits of nature and use them to decorate your castle. Anything goes here! A handful of tiny tree cones, pebbles, frozen leaves, and bits of bark can all add some extra flair to your ice castle. Have fun and be creative.

If you build your ice castle in the woods, or another location away from the house, you might have trouble finding it when the snow begins to melt. To tackle this potential problem, help your child make a nature flag with a tall stick and some leaves or bark. Use the flag to help mark the spot where the castle has been constructed so that you can stop by and visit, or perhaps build some additions, during the remaining days of winter. As you leave the construction site, take a moment to look back at your nature flag, taking note of the view so that it is easy to find the next time that you return from that direction. Looking back to see where you came from as you walk along a trail is a great safety skill to teach children, and this is a fun way to reinforce it.

When you have finished building your castle, you might want to spend some time breaking apart any leftover icicles to look at the core of the ice that makes them. Look for observable layers in the ice that give a picture of how the icicle slowly formed as water dripped downward a little at a time.

For another fun twist on this project, you can try adding food coloring or natural dyes (such as beet juice) to your icicles when you make them. By using a variety of colors, you can make a multicolored castle that will sparkle with vivid color in the winter sun.

BUILD YOUR OWN ICICLES

Sometimes nature only partially cooperates with the desire to build a fairy castle out of ice; perhaps providing the cold but not the building materials themselves. In that case, there are ways you and your kids can make icicles on your own and then plan a building day when the weather is cold enough for successful construction.

Option 1: FOR ADVANCED ICICLE BUILDERS

This method is ideal for older kids or for parents who suspect it might be easiest to do this part of the project ahead of time.

WHAT YOU WILL NEED

- WAX PAPER
- DUCT TAPE
- SCISSORS
- LARGE POT OR OTHER CONTAINER TO HOLD THE ICICLE MOLDS WHILE THEY FREEZE

WHAT YOU WILL DO

Cut a square of wax paper approximately 10″ × 12″. Cut larger or smaller squares to make icicles of differing sizes, keeping in mind the size of the freezer space if freezing indoors. Cut off one corner on the shorter side of the square at about a 45-degree angle. Starting at the cut corner, roll the wax paper to create a cone. Secure with tape.

With a strip of duct tape, starting at the bottom of the cone, wrap the wax paper cone at an angle, adding more tape as needed until you reach the top of the cone. Pinch the tape at the bottom of the cone and press any seams to be sure they are watertight. If the mold leaks, press the duct tape to secure it better and add more duct tape to "repair" leaks as needed. This may take a bit of practice, but it is fairly entertaining to test the molds for watertightness and repair spots that spring a leak.

Take a pencil or chopstick to smooth the wax paper on the inside of the cone, pushing it on the inside to make sure the duct tape is secure.

Fill the cone with water and stand it upright in the pot (if you have some snow available, pack it in around the icicles to help stabilize them as they freeze). Place in the freezer or outside in freezing temperatures. Check the molds regularly as they freeze, and add water if the water level has dropped.

Once it's frozen solid, run the cone under warm water to release the icicle. The wax paper may come out with the first icicle, leaving the duct tape mold intact. If it is watertight without the wax paper it can then be used to make more icicles (patch as needed).

Create as many molds as you like to build up a stockpile of icicles for castle building.

Option 2: ICICLE BUILDING FOR THE YOUNGER SET

This method is super easy and can be undertaken by the youngest of builders.

WHAT YOU WILL NEED

- POPSICLE MOLDS

WHAT YOU WILL DO

Fill popsicle molds with water and set them outside overnight or put them in the freezer, leaving out the sticks.

Once the water has frozen solid, pop the icicles out of the molds by briefly running them under warm water. Gather up your icicles and take them outside for a cool day of building. If you would like to make a larger number of icicles than your mold allows, simply work in batches. Once you have taken one batch of icicles out of the mold, place them in a paper lunch bag and store in the freezer while you make the next set. The paper bag should help to keep the icicles from sticking to one another.

For a colorful twist, and to brighten up all of that winter white, you can also have your child add a few drops of food coloring to the icicle water before it goes into the freezer.

WINTER BIRD BLIND

One of the most remarkable things about observing birds during the cold winter months is just how very comfortable they appear to be in even the worst of weather conditions. There isn't anything quite like watching a fat and fluffy male cardinal, unmistakably dressed in crimson, flitting around the yard in subfreezing temperatures and snowdrifts three feet tall. There is an undeniable charm to the ability of such tiny creatures to be so hardy, and winter can be an enjoyable time to bird-watch, particularly if you use the snow to your advantage.

To build a simple bird blind, kids can tug on a warm pair of waterproof mittens and begin piling snow up near a feeder or other location frequented by your neighborhood birds. Pile the snow until it is high enough that an interested party can pull up a lawn chair and sit behind the blind. Be sure to position the blind so that the birds are not able to people-watch but the people can easily enjoy a bit of bird-watching. For added entertainment (and comfort), come to the show outfitted with a thermos of hot chocolate and a colorful field guide for easy identification of any unknown backyard visitors.

Once your blind is in place, it is likely to stay put for most of the winter, provided that you make a few repairs as necessary and have cooperative cold temperatures. With this in mind, you may want to consider participating in the Great Backyard Bird Count hosted by the Cornell Lab of Ornithology each February. It is a fantastic opportunity to get the whole family involved not only with bird-watching but also with participating in a successful citizen science project. Kids will love feeling like official scientists in the field as they take notes about backyard visitors and submit their totals to the lab. More information about the Great Backyard Bird Count can be found at the following website: http://gbbc.birdcount.org.

DIY ICE STORM

Although heading outside during an actual ice storm isn't recommended for safety reasons, this project will help you and your child understand a little bit about how ice storms work. And, as an added bonus, it will also allow you to re-create the beauty of ice storms without the destruction and damage they are so famous for causing.

WHAT YOU WILL NEED

- NATURAL BUILDING SUPPLIES SUCH AS STICKS, TWIGS, AND LEAVES
- A VASE OR CONTAINER TO HOLD YOUR FOUND ITEMS ***(optional)***
- SPRAY BOTTLE FILLED WITH WATER
- FREEZING OUTDOOR TEMPERATURES

SCIENCE BEHIND THE SCENES: ***Freezing Rain***

The process through which rain becomes ice, freezing the landscape and dramatically changing the way the world looks and feels, is a meteorological event worth exploring. As you do, you and yours may well have questions about the weather phenomenon we call freezing rain. When a rainstorm arrives in the form of ice, or freezing rain, it is called an ice storm instead. Ice storms occur when there is a warmer layer of air above a very cold layer closer to the ground. The water is warmed as it falls, so it doesn't freeze into sleet but instead turns into ice when it hits surfaces closer to the ground that are at or below freezing temperatures. Ice storms are often destructive, and although the sparkle of the ice-coated landscape they create is beautiful, they can also leave a lot of damage behind.

WHAT YOU WILL DO

Arrange your sticks in a vase or use them to build a freestanding structure. Log-cabin-style buildings are fun, but let your imagination run wild. Play with different building types and arrange the sticks in a variety of ways to catch the water and allow for interesting ice formations to take shape.

Working outdoors in an area where the temperature is cold enough to freeze water, use the spray bottle to mist the sticks until the entire surface of each stick is thoroughly wet and drips of water begin to form. Let the first misting of water sit long enough for the water to become ice.

Continue to coat the arrangement or structure that you have made with water, letting each application freeze before adding a new one, and watch the layers of ice build up. More water can be added periodically during outside play time, but it can also be used as a motivation to get out the door multiple times during the course of a day. As long as the temperature outside stays below freezing, the ice will continue to build up on the structure and icicles may even begin to form! If the temperature is exceptionally cold, you will want to store the water bottle inside between uses.

Try weighing your sticks both before and after you spray them with water. How much more do the sticks weigh when they have been coated with ice? This is a good way to understand just how ice storms can wreak such havoc when they happen on a larger scale. Ice can be quite heavy!

You might also experiment using different materials in your ice storm. Although sticks are a great way to go, real ice storms affect everything that they touch. Try spraying natural objects with a variety of textures and shapes. Seed heads, winter berries, and leaves all make interesting materials to work with.

Create multiple ice storms around the yard in areas that receive varying amounts of sun or shade. How does the ice behave differently in areas that get more sunlight than others? Does it freeze or melt more quickly? Is the visual effect the same as in the shaded areas of the yard, or does the ice look different?

WINTER WHITES

Knowing that snow is made up of ice crystals, you may be curious why it is that snow is white, while ice is often clear or see-through. To understand this, it is important to know a bit about color and light.

The colors that we see in the world around us are really just the result of our eyes interpreting the different wavelengths of light. We see one wavelength of light as blue but other wavelengths as red or green. When light shines on snow, it cannot simply pass through, even though snow is made of water, which we think of as being clear. Each snowflake is a collection of individual ice crystals, and each of those crystals reflects light. So light shines into a snowflake from above, and is bounced all over the place within the flake, being reflected from one ice crystal to another. The result is that all of the colors of light from the visible spectrum get mixed up inside the snowflake, making it appear white to our eyes. If things (for instance, algae or dirt) get into the water that the snow is made from, then it may no longer appear white at all but will instead be the color of whatever has been mixed inside it!

ICE ORBS

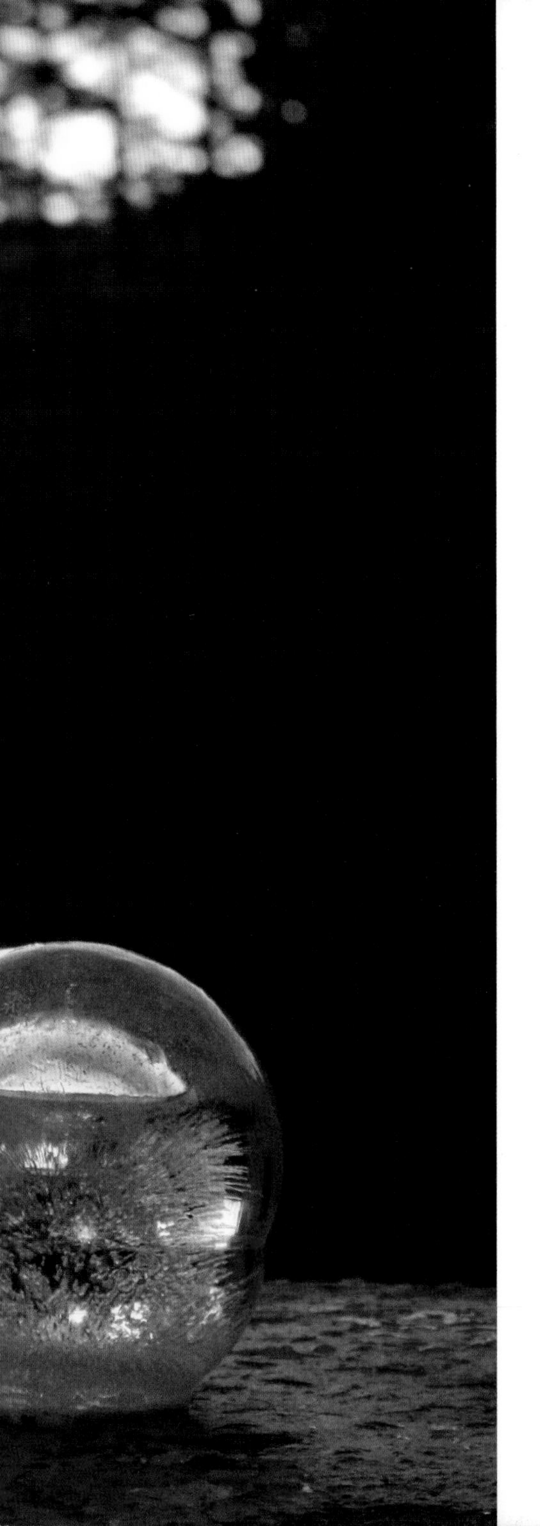

Ice orbs, sometimes called ice marbles, are an easy yet unexpected way to turn ice into playthings. Made from water frozen inside basic balloons, they can be piled into ice sculptures, rolled down sled runs, or set in the garden for a dash of winter color. Read on for instructions and you'll be freezing up your own set of these giant frozen orbs in no time!

WHAT YOU WILL NEED

- SMALL BOWLS ***(metal bowls work great)***
- BALLOONS
- COOKING OIL
- WATER
- FOOD COLORING ***(optional, but fun)***
- A FREEZING PLACE ***(outside in the winter or your freezer—the colder the better!)***

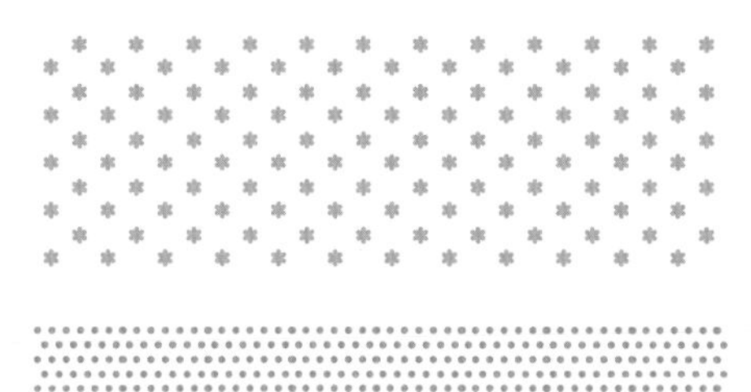

SCIENCE BEHIND THE SCENES: *Unusual Ice*

In the case of the ice orbs featured in this book, tools are employed to mold ice into a shape that is both interesting and potentially useful. This is also true when people make ice blocks for shelters, or when we fill an ice cube tray and toss it into the freezer. But what about the unusual shapes that ice sometimes takes without human intervention? When weather conditions are just right, water, temperature, and wind can become coconspirators in creating some unique ice formations. Ice can appear as sticks, curls, or even as pieces that are similar to bird feathers in their size and shape.

WHAT YOU WILL DO

Coat each bowl with a drop of cooking oil (this will help release the balloon after it is frozen).

If using food coloring, place a few drops of the coloring in the balloon (a little goes a long way and too much will stain clothes and hands). Then fill the balloon with water, making it a water balloon.

Tie the balloon and place it in the bowl with the knot down. Set the bowl outside (or in your freezer) to freeze. The time the water takes to freeze will depend on the temperatures and the size of the balloons, so a set of small balloons may freeze overnight if the temps are low, while a larger balloon may take a few days. Keep them out of the sun in a particularly cold spot if you will be leaving them for more than one night. This will keep the ice inside from melting in the sun during the days. Even if the outside temperature is very cold, the sun can be surprisingly strong!

Once the water inside is entirely frozen, peel away the balloon. You may want to employ a pair of scissors here. Use the orbs to build, play, and explore the properties of ice as it melts.

If you want to make lots of little ice orbs (and don't mind if they are a bit wonky) you can simply fill lots of small balloons and put them in a big pan together for freezing. These miniature orbs make wonderful building "blocks" that can be held together with snow and a spritz of water from a spray bottle. Simply hold them in place with water between the orbs until they freeze together.

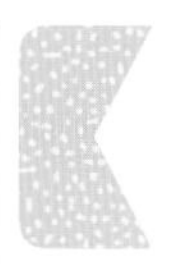

It can be fun to open up larger balloons before they are completely frozen to watch the water move around inside and to see the way the ice forms.

After they have been rolled down sled runs, used to mark obstacle courses, or hidden like Easter eggs in the snow, the ice orbs can be set out on display somewhere in the sun to enable you and yours to watch them melt over the course of the remaining days or weeks of winter. If they have been colored with food coloring, it might be also be interesting to take note of which colors melt faster than others, extending the science of this activity through a little chat with your kids about heat absorption.

You can also add natural materials to the inside of the balloons before filling them with water, freezing things like flowers, leaves, and berries into the orbs. As they melt, notice if the orbs with more materials mixed in tend to melt more or less quickly than others.

MAKING HAIL

Make weather on your kitchen counter! This simple activity may not be an exact representation of how hail forms in the atmosphere, particularly if you are using a rectangular-shaped ice cube tray, but it is easy, fun, and demonstrates the basic idea of water freezing in distinct layers. A perfect foray into weather exploration for even the youngest of scientists.

WHAT YOU WILL NEED

- ICE CUBE TRAYS
- TAP WATER
- FOOD COLORING
- YOUR FREEZER OR A VERY COLD FRONT PORCH

SCIENCE BEHIND THE SCENES: *Here Comes Hail*

Hail is one of the more fascinating forms of precipitation. Although you might expect the explanation of how and why hail forms to be complex, the process through which it forms is actually a simple one. It just happens to have amazing results.

When ice forms on a dust particle in the air and begins to move around inside a cloud, an ice pellet is formed. This ice pellet moves up and down on the wind drafts present within the cloud, and as it moves more and more layers of ice are added to the pellet. Eventually it forms a hailstone. The hailstone will continue to grow with each trip through the cloud, until it is too large to be pushed up by the draft and falls to the ground. The more powerful the wind pushing the hail up into the cloud, the bigger the hail will become before falling. There have been some record-breaking hailstones over the years, including some that fell in the United States that were nearly eight inches across!

WHAT YOU WILL DO

Begin by mixing a few drops of food coloring into about ¼ cup of water. Fill your ice cube molds up about a quarter of the way with the colored water. Freeze until solid.

Mix a second color of food coloring into another ¼ cup of cold water, and pour this new color of water on top of the frozen layer already in the ice cube tray, filling the compartments a total of half full. Freeze again.

Repeat this process until you have filled the ice cube compartments and each cube has layers of four or more colors.

When the cubes are totally frozen, remove them from the tray and observe. The layers that have formed are reminiscent of the layers of frozen water that collect to form hail!

If you'd like to make super-sized hail, you can easily do so by using containers of different sizes, or even shapes, to make your hail pellets.

Making hail in a variety of sizes can also be an interesting way to play around with how falling hail can impact, or even cause damage to, the world around us. Find a soft spot of earth somewhere outside, and try dropping hail pellets of different sizes onto the ground there with different levels of force or from different heights. What happens?

BECOME A SNOWFLAKE **SCIENTIST**

This is one of those activities that is remarkably simple to execute without sacrificing any of the "wow factor" that accompanies a truly great scientific observation. Taking a close look at individual snow crystals is a particularly cool way to help kids make connections between weather conditions and the resulting phenomena that they can see and touch. And, this activity also extends itself nicely into artistic endeavors. On an especially cold day, running back inside to sketch snowflakes next to the fireplace can be just as pleasant as looking at the real articles up close.

A little tip before we begin: we have felted and cut off the arms of a dark-colored, extra-large wool sweater for this activity. (To felt a wool sweater, simply put it through the washing machine.) It is great to keep these "arms" in the freezer: that way they are already chilled and ready for the perfect snowy day. They are easy for kids to slip on over their snow jackets, and this prevents the inevitable dropping of a loose bit of fabric handled by bulky mittens.

WHAT YOU WILL NEED

- A PIECE OF DARK-COLORED FABRIC OR FELT
- A SNOWY DAY THAT ISN'T TOO WINDY

SCIENCE BEHIND THE SCENES: ***No Two Alike***

You may have heard the popular adage that "no two snowflakes are alike." This is true for a handful of reasons, including the fact that as snowflakes fall, they often end up touching neighboring snowflakes or other objects and this can make changes to their original shape. What you may not know is that the shape of a snowflake, also called a snow crystal, can tell you what temperature it is outside. One of the most amazing things about snowflakes is also one of the things that a lot of people have never even heard: snowflake shapes are determined by the air temperature! Most snowflakes are six-sided, but the arrangement of those sides varies. The sides can branch out like a tree (dendrite snowflakes) or they can look a little bit like flower petals (sectored plate snowflakes). See the Snowflake Temperature Chart opposite.

WHAT YOU WILL DO

To use snowflakes to tell you about the temperature outdoors, you first need to bundle up and head out into the falling snow. We have always found that the perfect snow for this type of exploration is the soft, fluffy stuff that isn't too wet or slushy.

Once you have a proper snowstorm to wander into, capture individual flakes by laying out a piece of dark-colored fabric or another surface that will provide contrast and make the snowflakes visible. Look closely at the flakes, and reference the snowflake chart below to see if you can identify which of the five types of snow you have captured. When your fabric fills up, simply shake off the flakes that are occupying the space and begin again. Finally, take the outside temperature to see if it provides a hint about what type of snowflake you are holding.

SNOWFLAKE TEMPERATURE CHART

TEMPERATURE RANGE *(Degrees Fahrenheit)*	FLAKE TYPE
25 to 32	thin plates
14 to 25	columns
10 to 14	plates
3 to 10	dendrites
-8 to 3	plates

While you do not know the temperature at which they formed, can you identify the flakes in the following photos?

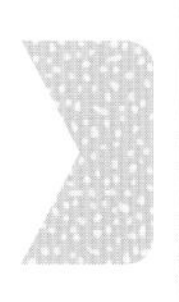

MORE TO EXPLORE

A simple pocket microscope can be of fantastic use here. Look for a model that will easily allow you to magnify the size of the snowflake without having to disturb its location on the fabric.

You may not be able to keep them in a palm or pocket, but with the right tricks up your sleeve, you can collect snowflakes with a camera. Not all snowstorms bring the best conditions for photographing snowflakes. Sometimes snowflakes are clumped together in a slushy mess, or melt just as they touch the ground. Other times it is just too cold and windy to capture those pretty flakes. But when the flakes are coming down fluffy and perfectly formed, the wind is calm, and the temperature is just right, it is a great time to take the kids out for a snowflake hunt, camera in tow, to

WONDERING ABOUT OUR WORLD

WILSON "SNOWFLAKE" BENTLEY

Of the artists known for using nature as a medium, Wilson "Snowflake" Bentley stands out for his incredible perseverance and attention to detail. Born in Vermont in 1865, Snowflake Bentley spent most of his life refining a process for being able to photograph snowflakes in the fleeting moments of their perfection, just after falling, before they quickly begin to melt away.

A dedicated student of the natural world, Snowflake Bentley spent years engaged in an intense trial-and-error process, hoping to find a way to photograph single snow crystals. Using both a microscope and a bellows camera, Bentley was finally able to produce the first known photograph of a single snowflake in 1885.

By the time that Bentley passed away in 1931, he had taken over five thousand photographs of single snowflakes. He did not find any two alike.

These days, visitors to the Snowflake Bentley Museum in Vermont can view many of the photographs taken by William Bentley during his life, getting a firsthand look at the amazing detail that he was able to capture with fairly basic technology. For those who don't have a trip to Vermont on the calendar anytime soon, a trip to the local library to pick up a book about Bentley would be a fine substitute. ***Snowflake Bentley*** by Jacqueline Briggs Martin and ***Snowflakes in Photographs*** by Bentley himself are both excellent choices for a snowy afternoon read by the fire.

photograph the snowflakes. Since it takes a steady hand, an adult might want to take the photos of the snowflake discoveries using the tips here to capture them in their best possible light, or help the involved kids with getting basics down before sending them off to try it on their own.

HINTS FOR GREAT SNOWFLAKE PHOTOS

Get up close and look at the snow from many angles. You might even need to get down on the ground, so go ahead and wear your snow gear.

Some cameras can be set to a macro or close-up setting. If your camera has this setting, use it.

Try to take pictures with something dark in the background. You can bring out a bit of black felt you have kept in the freezer (to keep it cold) or find dark backgrounds in nature. Tree bark, dark brown dirt, green moss, or even rose hips make great natural backgrounds for snowflake hunting.

Crop your photo in a photo-processing program on your computer. Most cameras can't get close enough for lots of detail. As long as the photo is in focus, cropping it can help show off the snowflake.

In the photo-processing program, change the contrast and play with other settings to help highlight your snowflake.

Have your kids use the pictures to make a collage or a holiday card. Share them with friends and family.

Use what you have learned about the different shapes of snowflakes to identify your flakes and predict the temperature outside.

FROST PAINTING

Frost provides unique and interesting texture to items in the natural world, turning ordinary objects into something altogether different. A frosty, cold morning is the perfect time to head outside for frost painting. The supplies needed are minimal, but the results are surprising and wonderful.

WHAT YOU WILL NEED

- PAINTBRUSHES OF VARIOUS SIZES AND SHAPES
- A CUP OF WARM WATER
- A FROSTY SURFACE

SCIENCE BEHIND THE SCENES: ***First Frost***

Upon the arrival of the first frost one chilly fall morning, your child may pose a few questions about the white sparkly dust blanketing the ground. What makes frost? Why does it form on some things and not others?

Frost is water vapor in the air that has condensed and formed ice on a surface. Some surfaces cool faster than others, so frost forms on those surfaces first. For example, frost-covered grass that is out in the open may be at or below 32 degrees, even if the air temperature is not freezing, while grass under a tree has not lost enough heat overnight to reach freezing, so will not be covered in frost. If the air temperature is extremely cold, the air will not hold much water vapor and frost will not form at all.

WHAT YOU WILL DO

Look around for a frosty surface to "paint" on. After going on a frost walk your child is likely to be keen to identify spots with frost that would make a great canvas for a painting.

Once you find some good frost, begin using fingers, paintbrushes, and water to alter the surface of the frost. Different designs can be created by simply moving the frost around, or by dipping the brush in water to melt the frost.

Various surfaces will lend themselves to either abstract designs or more "realistic" art depending on the texture and type of frost that has formed there.

As the sun comes up keep an eye on the art to watch how it changes as the frost melts.

If you have a sensitive little one be sure to prepare them ahead of time for the idea that their art is going to melt and become water again!

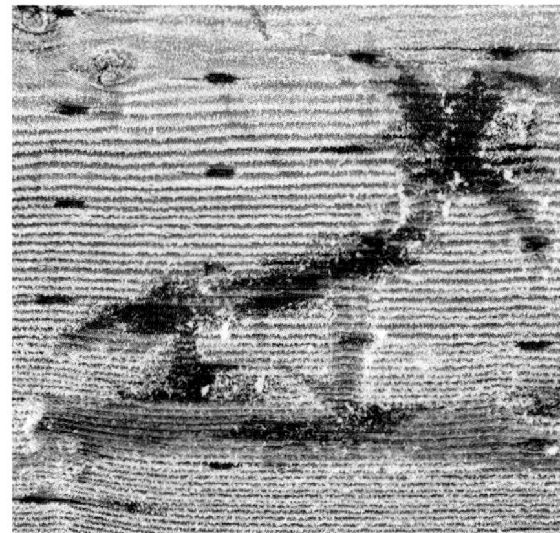

When the weather forecast promises that frost is likely, choose a handful of items to set outside overnight. Spend some time making predictions about which ones might collect the most frost, or what that frost might look like. In the morning, head out to check the accuracy of your predictions!

You can also get two or more of the same item and try placing them in different spots in the yard. On a frosty morning, visit the yard to see which of the objects have frost on them, and think about how their placement in the yard might have impacted their ability to become frosty or not.

Monitor the temperature in shady or sunny spots around the yard with a small thermometer. How does the temperature in the frosty areas of the yard differ from the temperature in places in the yard with less frost?

EXPLORING FROST

TAKE A **DISCOVERY WALK**

When late autumn arrives each year, the appearance of an early morning frost is often the first sign that winter is right around the corner. The nighttime temperature starts to drop, hats and mittens are found for chilly mornings, and the world outside glitters with the sparkle of frost on leaves, grass, garden boxes, and cars. As our bodies adjust to the new cooler weather, it can take a little extra effort to get out the door each day. But a frost walk can be a great reason to bundle up with a few extra layers and head outside to get a closer look at what is making the world so sparkly and bright.

Whether you are walking around your own backyard or down to the local park, there are fun things to investigate and questions to ponder along the way.

Here are a few questions to ask your child as you go:

Has frost formed on everything? Why? Why not?

How does it take shape on different surfaces? Is it a thin sheet or spikes standing up on the surface of an object?

Is it "sharp" or fragile?

Is it different on raised surfaces, like the veins of a leaf, or flat surfaces, like a car window?

Where does the frost melt first? Sun? Shade? Dark surfaces? Light surfaces? Why?

HOMEMADE
ICEBERGS

If you walk near waterways in the winter you will often spot ice that has formed in layers along the water's edge. Come spring, you may even see some pieces of ice that have started to break off into the moving water and float downstream. While these are not considered icebergs, they behave in much the same way as an iceberg does. If a trip to the open waters in the Arctic North or Antarctic South is not in your vacation plans this year, you can still see icebergs in action in a nearby stream, or right in your own backyard, with the fun activity that follows. And since kids naturally love to throw rocks, or blocks of ice, into moving water, this is a natural extension of something they already have an innate drive to do.

WHAT YOU WILL NEED

- SNOW OR ICE
- A STREAM, LAKE, OR LARGE POT OF WATER
- GLOVES OR MITTENS TO KEEP YOUR HANDS WARM

SCIENCE BEHIND THE SCENES:
The Origins of Icebergs

Icebergs are generally associated with the icy cold sea waters around the Earth's poles, and most people know at least a little bit about them. But your young explorer may have more involved questions about icebergs and how they are formed.

When glaciers move toward the ocean, large sections of ice break off and fall into the water, forming icebergs. Because icebergs were once part of a glacier, they are made up of fresh water. Often sections of an iceberg look blue; the extreme density of the ice causes it to absorb all other colors on the color spectrum and reflect blue. Most icebergs are found near glaciers. But the wind and ocean currents can also carry them out into the open ocean. The International Ice Patrol was formed to monitor iceberg movement after the ***Titanic*** struck an iceberg and sank in the North Atlantic ocean in 1912.

WHAT YOU WILL DO

Have your kids pack damp snow into a ball, adding layers to build it up, just like they do when getting ready for a snowball fight. Alternatively, use a piece of ice that is large enough to see well in the water.

Gently place the snowball or ice in the water, taking extra caution near any moving water if you happen to be along a stream, river, or lake. Watch the "iceberg" and make some observations about how it behaves in your makeshift ocean.

- DOES IT FLOAT OR SINK?
- HOW MUCH OF THE ICEBERG IS ABOVE WATER?
- DOES IT MOVE ON ITS OWN?
- HAS IT CHANGED COLOR IN ANY WAY?

You might also spend a bit of time talking over how icebergs are made, and how it is that they move around in the water.

If you are near salt water you might introduce the idea that salt lowers the freezing point of water and therefore will make an iceberg melt faster. This concept can be re-created at home with two bowls of water, one filled with fresh water and one filled with salt water. Take a few ice cubes, similar-sized pieces of ice from outside, or compacted snowballs and float the ice in the two bowls. Make observations about which one melts first.

You can try freezing salt water at home as well. The materials needed are few, and the resulting opportunity for observation is a nice way to bring the science of the sea to your kitchen counter.

Dissolve ½ teaspoon of salt in ½ cup of water. Fill an ice cube tray with fresh water on one side, and the salty water you have mixed up on the other side. You might want to put a piece of tape or something similar on one side of the tray to mark which water is which. Place the ice cube tray in the freezer. Set the timer for fifteen minutes.

After fifteen minutes have passed, check the ice by wiggling it just a little bit. Take notes about the state of the water in both the freshwater and saltwater compartments of the tray. Has any ice formed on the top of the cube? Does it feel solid or slushy?

Continue to check at regular intervals, approximately every fifteen minutes, and take notes about what you observe. Leave the cubes in the freezer overnight and make one last observation the next day. Remove the ice cubes from the tray to observe how the ice is formed, the way it feels, and note any differences in transparency. You can extend the activity further by leaving the ice cubes out to make observations about which ones melt faster.

ICE AT SEA

In places where the winters are long and cold, the temperature can drop low enough to freeze the salty sea. But sea ice is not simply frozen salt water. Salt and water can both freeze, but they don't freeze at the same temperature. Salt freezes at much lower temperatures than water. As the water freezes, ice crystals form, but the salt does not become part of the crystal because the temperature is still not cold enough to freeze the salt. Therefore sea ice is almost pure fresh water; just like glaciers.

The newly formed ice crystals are small and needle-like. They are called "frazil." As more and more frazil crystals are formed they float to the surface of the water and start to cling together. Ice sheets start to form as the frazil accumulates.

Sea ice is formed in different ways depending on the state of the water, the location, and changes in weather such as wind and temperature. All of these factors together create many different types of ice that move and change with the seasons.

2

SAVING FOR A RAINY DAY

There is much to be said for the cozy atmosphere created by simply staying indoors with a warm beverage and a good book on a wet and windy day. Well, for the grown-up set anyway. For kids, a long rainy day can be less of a reprieve from the bustle of everyday activities and more of a burden. This is especially true when there are many days of rain in a row. The good news is this: rainy days are full of opportunities for learning about weather in our world, and rainfall can actually provide us with fun and unusual means for making great art.

Rain is the most common type of precipitation. It is created when water droplets form inside clouds, growing larger and larger until they are too heavy for the cloud to hold. When this happens, the droplets fall as rain. For this to occur the droplet must have enough weight to fall through the updraft, or the air keeping the cloud up. Droplets grow by bumping into and joining with other droplets. They grow into raindrops, which can then collide with other raindrops and grow even bigger, a process called "coalescence." As they begin to fall through the cloud they quickly join other drops and droplets to grow rapidly until they reach the surface of the Earth as rain.

Rain can sometimes begin as ice high up inside a cloud. As the ice, or snowflake, begins to fall, it may encounter warmer air below. When this happens, the ice melts and becomes a raindrop. It may join with other raindrops on the way down to the ground. This is one of the reasons why it can snow in the mountains nearby, but it can be raining down at lower elevations where the air temperature is warmer.

Once rain reaches us here on the Earth's surface it becomes an ideal medium for art, science, and experimentation. In the pages that follow, you will find simple and entertaining activities that allow for exploration of falling rain, forming puddles, and other wet weather phenomena as it comes your way.

Think of this chapter as your manual to making the very most of a rainy day.

PUDDLE
EVAPORATION
ART
Crayola

With all of the potential for splashing around and kicking about, puddles can be an endless source of fun on a rainy day. After the rain stops, tracking the evaporation of a puddle is an entertaining investigation, and if done just so, can allow for the creation of some interesting artwork. Follow these instructions to turn a disappearing puddle into a unique display of color and line that can be enjoyed until the next time the rain falls.

WHAT YOU WILL NEED

- A PUDDLE LOCATED ON A SOMEWHAT FLAT SURFACE ***(such as concrete or asphalt)***
- SIDEWALK CHALK ***(multiple colors are fun, although not necessary)***
- TIME TO RETURN TO THE PUDDLE REGULARLY ***(the evaporation rate will vary depending on your climate)***

SCIENCE BEHIND THE SCENES: *Evaporation*

Have you ever noticed water outside somewhere, and watched as it disappeared over time? Maybe the sidewalk quickly dried up after a rainstorm, or a puddle got smaller and smaller as the sun's heat warmed the day. Many inquisitive young naturalists will have questions about where the water has gone. That water is not really disappearing but is changing state, from liquid to gas, through the process of evaporation. As we learned in the review of the water cycle in the introduction to this book, evaporation is an important part of the water cycle and is one way that water that has fallen to Earth through precipitation is able to return to the atmosphere to start the cycle again.

WHAT YOU WILL DO

While it is raining, search for an interesting puddle that has formed in a location where sidewalk chalk can be used. Get creative, looking on boulders and other surfaces. It is helpful if it is close to home where it can be checked regularly.

While out puddle-hunting ask some questions about puddle formation and really get up close and personal with your puddle.

- WHERE HAVE PUDDLES FORMED?
- WHY HASN'T THE WATER SIMPLY DRAINED AWAY?
- WHAT IS KEEPING THE WATER THERE TO FORM THE PUDDLE?
- DOES A PUDDLE FORM IN THE SAME SPOT EVERY TIME IT RAINS?

Once the area around the puddle starts to dry, have your child outline the puddle with chalk. Use a brightly colored piece of chalk that will contrast well with the ground surrounding the puddle.

Check the puddle at regular intervals to make a new outline as it starts to evaporate. The puddle will evaporate more quickly as it gets smaller and has less water in it, so check more and more frequently as the puddle shrinks.

Talk about why it is that the puddle is getting smaller and discuss where the water could be going. Remember the water-cycle basics in the introduction. They are there if you need them! Once the puddle has completely vanished, you will have a wonderful rainbow outlining the life of your puddle.

If there are still clouds in the sky when your puddle art is being created, point them out and discuss how the water is entering the air again to become clouds.

Think about snapping a few photos to hang on the fridge to create a more permanent piece of artwork from your puddle observations. These can be a continued source of prompts for questions about evaporation and the water cycle.

Track the growth of a puddle with pebbles while it is forming during the rain. To do this, place a line of pebbles or rocks along the edge of an existing puddle to outline its shape. As the puddle grows, add another rock "ring" surrounding it. Later, when the puddle evaporates, it will be especially interesting to see how large it once was and to think about the phases of its growth.

Make your own puddle by finding a spot to dig in your yard. Look back at the questions about puddles to help guide your thinking

about whether or not your spot will hold water when the rains come. Some low spots do not hold water because the water percolates, or seeps into the ground to become groundwater, instead of evaporating into the air. There are many variables, but compaction (how much the soil is pushed together) and the type of soil (like sand or clay) greatly impact whether or not a spot will hold water. Wait for a rainy day to see if your spot will become a puddle, or if no rain is in the forecast you can test your puddle location out with a garden hose. What happened? Why?

ANDY GOLDSWORTHY

Born in England in 1956, Andy Goldsworthy is famous for utilizing both unusual methods and materials in the artwork that he creates. Goldsworthy uses natural objects almost exclusively, and his art generally comprises leaves, sticks, ice, snow, stone, or flower petals. Often using "tools" such as his own teeth and fingers, Goldsworthy tears, breaks, or weaves together natural objects to create pieces of art that demonstrate the beauty of nature, while also reminding us of the ways in which it is always changing.

Goldsworthy creates much of his artwork within the natural environment itself, leaving it behind after it is complete to bravely let nature reclaim the materials that he worked so long and hard to turn into art. Goldsworthy often photographs his work, both in its finished form and as it falls apart over time. His work is a reminder that the joy of creating is in the process itself, not just the product—an excellent lesson for young artists everywhere.

Andy Goldsworthy has published a number of books containing photographs of his artwork, as well as the process through which he creates his pieces. There is also a documentary, ***Rivers and Tides***, about Goldsworthy that even younger children are likely to find fascinating and inspirational.

PAINTING WITH
RAINDROPS

When the weather turns wild, with the wind howling and rain collecting in puddles, it is easy to find yourself running from building to building trying to stay dry. This simple activity is a great reminder of the way that the presence of water in our world allows us to do more than just the ordinary loads of laundry and after-dinner dishes we know it is so useful for.

Yes, the water cycle is responsible for the greening of our grass, but it also provides the water needed for various art forms, many of them favorites among the small set. Watercolor painting is certainly among them. For added interest, this project takes the usual techniques employed in painting and turns them on their head just a bit.

Read on to find out how to get the rain to not only provide the water needed for a watercolor masterpiece, but how to get those fat, falling drops to do the work of painting for you.

WHAT YOU WILL NEED

- WATERCOLOR PAPER
- WATERCOLOR PAINT TUBES
- PAINTBRUSHES, STICKS, LONG GRASSES, OR OTHER PAINTING TOOLS ***(see page 84 for ideas on making paintbrushes from nature)***

SCIENCE BEHIND THE SCENES: ***Cohesion***

Looking at an ocean, or even a pond or puddle, the idea that every body of water is composed of individual drops of water seems almost farfetched. For the young, it can be difficult to imagine that something that seems like a whole is indeed a sum of many, many tiny parts, and that each drop of water can exist outside the larger body on its very own. Water is able to do this because of something called "cohesion." Cohesion, simply put, is the tendency of water molecules to be attracted to other water molecules. And despite the small size of a single drop of water, its cohesive properties are powerful. In fact, cohesion is even stronger than gravity, which is what allows water molecules to stick together in drop form rather than lying only side by side.

WHAT YOU WILL DO

Take the supplies outside on a rainy day, and find a place to set up where there is enough cover to begin the project without everything getting wet before you are ready.

If you are painting with a younger child, have him or her choose some paint colors they would like to work with and ask them where they would like the paint on their paper. Place dabs of paint on the watercolor paper as directed. Older children can add paint to their paper on their own. If they have not worked with tube watercolors in the past, let them know that a little bit goes a long way.

Lay the paper out where it will be hit by the rain. If there is a little wind use a few rocks to weigh it down. This will allow the paper to get wet and start to move the paint.

Once the paper is sufficiently wet, let your child use brushes, sticks, grasses, or other painting tools of their choice to move the paint around.

When he or she is happy with the painting and feels as though it is complete, place it somewhere flat and protected to dry, or leave it out in the rain and watch how it changes as the water dilutes the paint.

There will undoubtedly be requests for more. Keeping extra watercolor paper in a waterproof bag will help keep it at the ready for more rainy-day artwork.

MORE TO EXPLORE

Play with paint placement and colors, adding more paint as needed.

Cover a piece of paper completely with a fresh, thick coat of watercolor paint, then let the raindrops hit the paint to make spots and drips as the paint moves around.

If you only have a watercolor palette with cake watercolors, simply place your paints out in the rain for a bit to fill with water, let your paper get wet in the rain, and paint away as usual.

Drop water droplets into a glass or plastic container. Note the size and shape of each droplet, then have your child move the container to see if the water droplets will move and connect with each other. This connection is cohesion at work! To help reinforce the concept, you may want to add color to the droplets using food coloring. As the drops combine they will create a new color!

WONDERING ABOUT OUR WORLD

STAYING SAFE ON A RAINY DAY

Because storms are generated in the atmosphere far above us, it can be tricky to tell how near or far away an approaching set of storm clouds really is. But when thunder and lightning are present, there are ways to make a pretty good guess! To estimate how far away a storm is from where you are, simply start counting the seconds as soon as you see a flash of lightning. Stop counting when you hear the thunder. Every five seconds equals approximately 1 mile of distance (1.6 km) between the storm and where you are standing. If you can hear the rumble of thunder, that means lightning is nearby, even if you can't see it. It is best to watch the storm from indoors where you are safe from any hazards related to the weather conditions. (Learn more about storm safety in the preparedness guide titled "Thunderstorms, Tornadoes, Lightning: Nature's Most Violent Storms," published by the National Weather Service and available at www.lightningsafety.noaa.gov/resources/ttl6-10.pdf.)

NATURE PAINTBRUSHES

When it comes to using ordinary things for extraordinary purposes, this project is a favorite. The wet and windy days of spring or autumn are an ideal time to get out and gather twigs blown to the ground by seasonal gusts. Add a handful of soft wet leaves and a piece of twine, and you'll have a handy brush ready for watercolor painting in no time at all.

The artwork resulting from the use of these brushes is sure to be unique, and the making of the brushes themselves provides a nice opportunity to explore the characteristics of individual plants and to consider all of the interesting adaptations that nature comes up with to allow plants to thrive in a variety of conditions. Young botanists and aspiring artists alike are sure to find something to love when it comes to this activity.

WHAT YOU WILL NEED

- STICKS FOR THE BRUSH HANDLES
- LEAVES, GRASSES, OR ANY OTHER NATURAL MATERIALS YOU WANT TO USE AS BRISTLES
- TWINE OR STRING
- A HELPER WHO CAN PROVIDE A SECOND SET OF HANDS AS NEEDED

SCIENCE BEHIND THE SCENES: *Plant Adaptations*

Like other living organisms, plants and trees adapt to their environment over time. They develop special features that allow them to take advantage of the natural resources available in the areas where they live, a fact that can result in plants having leaves with a wide variety of interesting textures, shapes, and even colors. When exploring the myriad of ways plants are different, you might want to point out to your child some of the many variations found in the leaves you have right in your own backyard. Some plants have hairy or fuzzy leaves to help them retain water or to protect them from insects looking for a meal. Others have leaves with sharp or rough edges to discourage deer or other hungry forest animals. Thick, waxy leaves can help a plant retain water, while small leaves or needles can prevent a plant or tree from getting overheated.

WHAT YOU WILL DO

Begin by choosing a strong and relatively straight stick or twig to use for a brush handle. If the plan is to make a number of brushes, it is fun to choose sticks in a variety of lengths to allow for holding the brush tip different distances from your body when painting later.

Lay down a length of twine about five inches long and place a small bundle of your chosen brush material on top of the twine in the center of the length. Then lay the stick in the center of the bundle and add more brush material on top of the stick. Let the bundle overlap the top of the stick by about a half inch or so.

Have your helper hold the bundle of brush material together, holding the stick firmly in the middle of the bundle.

Tie the twine with a good solid knot to secure the stick and the brush material together, making sure that the brush material is firmly attached to the brush handle. Then wrap the twine around the stick a few times, securing the brush material in place at the end, and tie off.

Dip your brush in the paint of your choosing and use it to create something wonderful! You can also place a dot of tube watercolor paint directly on the paper and let the artist move the paint around right on the page.

Make a whole set of brushes using a variety of materials for the bristles of the brush. Experiment with the different lines and shapes your paintings gain from doing this.

These brushes would also make a lovely gift for a friend. Package them up in a brown paper package with watercolor paper and a set of new paints, and you will have a rustic gift for giving that any outdoor enthusiast or budding artist will appreciate. You might even add a small amount of extra twine to the package so that your friend can add fresh pieces of greenery to the brushes after they experiment with painting with the dried versions you've sent.

CATCHING RAINDROPS

As rain finds natural resting places on twigs, cones, moss, or leaves, the drops look like little jewels hanging down, decorating a gray, dreary day with the sparkle of reflected light. The process of collecting those individual drops of rain is both incredibly simple and surprisingly fun. Follow the directions below to gather the fallen drops of rain in your own yard or favorite park. While you work, think about how many it would take to make a sip, if you really were in need of a drink, or speculate about how many drops it would take to fill a pond or even the ocean.

WHAT YOU WILL NEED

- A SMALL (PREFERABLY CLEAR) CONTAINER
- CURIOSITY AND A LITTLE PATIENCE

SCIENCE BEHIND THE SCENES: ***Collecting Water***

Although human beings collect water in any number of containers for any number of reasons, the process looks a little different in the natural world. But collection is very much an important part of the water cycle, and it deserves at least a bit of attention here. Even the smallest weather watcher may wonder where water goes once it falls back to the Earth's surface. Water can collect in different places depending on where it falls when it returns to the surface on the Earth. Sometimes precipitation falls directly back into a water source like the oceans or lakes, rivers, and streams. These water sources are called "surface water" because they are on the surface of the Earth. Water can also fall on land and either soak into the ground to become "groundwater," or it can run off into lakes, rivers, streams, or the ocean. Groundwater is water that is held underground.

WHAT YOU WILL DO

Head outside on a rainy day to observe the raindrops gathering on different surfaces. Note the size and shape of the drops that you see, and observe how they reflect the world around like a mirror. A few things to think about along the way:

- WHERE DO RAINDROPS GATHER?
- ARE THEY ALL THE SAME SIZE?
- DO THEY TAKE DIFFERENT SHAPES DEPENDING ON WHERE THEY ARE HANGING?

Start collecting drops by gently placing your container under the drop to catch it, or by shaking a branch or twig with the container underneath. You can also encourage your child to simply hold the container out to see if any raindrops fall in. You may want to ask a few more questions at this time.

- HOW OFTEN DOES A DROP FALL IN THE CONTAINER DIRECTLY FROM THE SKY?
- WHAT IMPACTS THE NUMBER OF DROPS COMING DOWN AT ANY ONE TIME?

If you would like to, you can count the raindrops as they fall into the container to keep track of your drop collection as it grows.

Continue collecting as long as you like. Note how much water has accumulated in the bottom of your container from time to time.

Try using containers of different shapes and sizes to collect rainwater.

Place a funnel in one container to collect water and compare it to a container the same size. Which one collected more water? A measuring cup is handy for aspiring meteorologists who would like more exact measurements for the amount of water collected.

Store your collected rainwater and use it for water-color projects at another time. A small mason jar with a tight-fitting lid works nicely. You can even label the jar with the date you collected the rainwater and where it came from and then add this information to the signature on the painting it is used for.

If the measuring cup isn't quite satisfactory for your serious young scientist, consider going all in and making a rain gauge to use for measuring rainfall around your yard. Place a ruler inside a jar or other container so that it stands up straight against the wall of the container. If need be, anchor it in place with some strong tape. A straight ruler will increase your ability to get an accurate measurement from your rain gauge.

Leave the container outside in an open location where any rain that falls can accumulate. You want to avoid setting it under gutters, eaves, or other areas that could result in gathering more or less rain than what is actually falling on open ground. Note the time when the rain begins, and when you stop letting rain fall into the jar, or when the rain stops on its own.

To see how much rain fell during the period of time when you were measuring it, take the ruler out of the container very carefully, and note the line where the ruler goes from wet to dry. That line tells you how much rain fell during your period of measurement. If your child is concerned that it will be difficult to see the line between wet and dry on the ruler, you could also put a couple of drops of food coloring into the rain water in the jar before taking the ruler out. This way, the water on the ruler will have some color in it and may be easier to see.

Tree roots grow underground within the tree's drip line. This means that the roots of most trees do not grow beyond the reach of the branches of the tree, where water will drip down from the leaves. Go outside on a rainy day and watch how leaves catch water and direct it to the ground below, right to the roots of the tree.

WONDERING ABOUT OUR WORLD

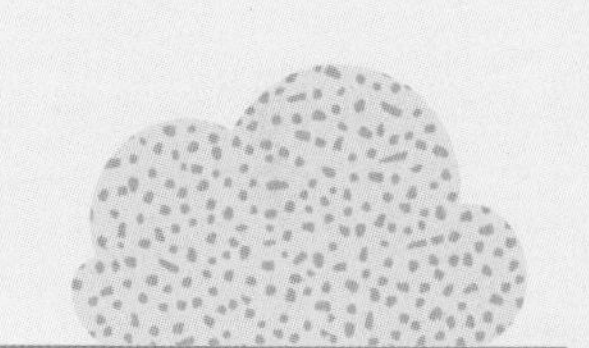

WHAT IS A METEOROLOGIST?

A meteorologist is a special kind of scientist who studies the atmosphere and the phenomena that take place there, including the weather. Meteorologists often work for the government or for other public agencies, but some also work for television or radio stations, providing weather reports to the general public. Meteorologists have made significant progress in their ability to predict the weather since computers have become widely available, and weather data is now available to people around the globe.

Meteorologists use a number of interesting tools and methods in their quest to accurately report and predict weather conditions, including instruments such as thermometers, barometers, and anemometers (a tool for measuring wind speed).

If you live near a television station you may consider calling to see if you can arrange a time for your child to visit with the local meteorologist. A rainy day is the perfect time to make a few phone calls and brainstorm some questions your child may have for the weather expert in your location. Your child may want to know about how the meteorologist became interested in studying the weather, what makes the weather in your area unique, or if there are geologic features close by that impact weather patterns. The more your child knows about the weather, the more questions they may have.

FLOATING
NATURE
GARLAND

This project is more than it may at first appear. Sure, it is a way to use flowers, fallen leaves, and other small pieces of nature to make a lovely garland to adorn a doorway or hang above a celebratory table. But, taken just a step further, it also becomes an experiment in the way that water flows and moves, acting as a constant force of change in the natural world. The way that water behaves and its impact on everything that it touches are always of interest to curious young naturalists. This project allows them to spend hands-on time thinking about how materials of different weight, shape, and size might act when tossed into a stream or pond, before taking the natural next step of finding out if their predictions were accurate.

WHAT YOU WILL NEED

- A VARIETY OF LEAVES AND FLOWERS OF DIFFERENT SIZES AND SHAPES
- STRING OR THREAD
- A HAND SEWING NEEDLE ***(it doesn't need to be very sharp)***

SCIENCE BEHIND THE SCENES:
Where Does Water Go?

As your family becomes more attuned to the patterns of rainy-day weather, questions may arise about the way water flows and where it goes. What makes water flow? Where is it going, anyway?

Like so many other things in our natural world, the flow of water in brooks, creeks, rivers, and streams is the result of one very important force of nature: gravity. Gravity is what causes water to move from higher altitude locations to places lower down, and it is the reason why water flows across the surface of our Earth. Water may move along the Earth's surface in a variety of pathways, but all of these flowing water sources follow the Earth's gravitational pull toward the center of the planet, or the equator. Flowing from higher ground to lower elevation results in our planet's streams and rivers finding their way to the oceans, which are at sea level.

WHAT YOU WILL DO

After a walk around your neighborhood park or through the backyard to gather supplies, lay out the leaves and flowers in a straight line. Ask your child to arrange them in the order he or she thinks will look best when they are strung together. (If it seems as though outdoor conditions might make this part of the project challenging, you could easily identify a nice place for floating your garland before going indoors to do your stitching.)

Thread the needle with a nice, long piece of thread or string, making sure that it is at least as long as the row of leaves and flowers that you have laid out. Tie a firm knot at the end of the thread. If you think that you will want to join the ends of the garland into a wreath, make sure to leave a long tail after the knot.

Starting at one end of the row of leaves, begin stringing them together. Push the needle through the front of the first leaf, somewhere near the middle. About a half inch or so from the first hole, push the needle back through the back side of the leaf so that it is now in front of the leaf again and there is a bar-shaped stitch on the back of the leaf. Gently pull the stitch taut, being careful not to tear the leaf, and repeat the stitching technique to attach the remaining leaves and flowers to the garland. Very young children can simply push the needle through and thread the leaves that way.

Once your garland is complete, tie off the other end of the thread or string. Then find a slow-moving stream or backyard brook to try floating your garland down. Local parks often have small bodies of water that work quite nicely for this sort of thing.

When you are done floating your garland, make sure to retrieve it from the water. This way, you will prevent any local wildlife from finding, and potentially eating, the string or thread that you used in stringing your garland. You can also make the thread extra long so it can be held as a tether to keep your garland from floating away completely.

If you'd like to make your leaf garland slightly more durable, you can coat the individual leaves with melted beeswax before stitching them together.

Once you have finished floating your garlands, try hanging them from trees around the yard, or even in various locations around the house. They'll provide a spot of color and visual interest, while also allowing for observation of how the plant materials change as they dry out.

NAME THAT STORM

Huge storms like hurricanes can cover many hundreds of miles and travel across multiple countries. In the past it was difficult for meteorologists to talk about storms with others around the world because the same storm could be called by many different names. Some storms were named after saints, others were given names of wives or girlfriends, still others were named after people the meteorologist did not like. In 1953, the U.S. National Weather Service started assigning women's names to hurricanes. Today, the names of both men and women are used. These days, the World Meteorological Organization (WMO) is responsible for giving names to hurricanes that occur in the Atlantic. They use a system in which there are six lists with the names of both women and men. The names on the list are in alphabetical order and use all of the letters of the alphabet with the exception of ***Q***, ***U***, ***X***, ***Y***, and ***Z***. The lists are used in a six-year rotation cycle. So every six years the same list is used again. If a storm, like Hurricane Katrina, causes a lot of damage, then the name is retired and never used again for another storm. A new name is chosen to replace it by the WMO. If all the names on a list are used up in one year, the letters of the Greek alphabet are used to name storms. Similar systems with different names are used to name storms that occur in the Eastern or Central North Pacific Oceans.

Giving proper names to the storms that work their way through your local skies can be a fun and creative method for helping kids to think about the weather conditions and how one rainy day differs from the next. This is especially good for morale during the long spells of wet weather that often come around during the months of autumn and early spring.

SURFACE-TENSION STRING ART

Surface tension is an undoubtedly useful natural phenomenon for any number of reasons. It is what allows a water strider to effortlessly move across the surface of a pond in search of a meal, and it is what causes raindrops to form. It turns out that it is also handy when it comes to fun science experiments and creating interesting artwork. This project provides an open-ended opportunity to play with the properties of water, to explore surface tension, and to create cool and colorful pieces of dynamic art at the same time. An ideal integration of science and art, to be sure.

WHAT YOU WILL NEED

- A TRAY OR PUDDLE FILLED WITH RAINWATER ***(if an evening storm is coming, set the tray out overnight to catch the rain)***
- YARN OR STRING ***(choose a variety of materials such as cotton and wool to explore how they might behave differently)***

SCIENCE BEHIND THE SCENES:
What Is Surface Tension?

Like most substances, water is made of molecules. The way that water behaves is directly related to the kind of molecules that make it. Many of you already know that water is also known by its chemical name, hydrogen oxide, abbreviated as H_2O. Surface tension is all about how those Hs and Os are attracted to one another. In a body of water, the hydrogen atoms (the Hs) are constantly trying to get cozy with the oxygen atoms of neighboring molecules (the Os). They can do this with molecules on every side of them, with the exception of those molecules on the very top, because there aren't any more water molecules there. This means that the water molecules on top are constantly being pulled down by the ones on the bottom, creating surface tension!

WHAT YOU WILL DO

Cut a length of yarn between one to three feet. Shorter lengths might be better for the younger crowd. Tie the yarn into a loop. If it's raining, tuck the yarn into a pocket or waterproof bag where it will stay dry until you get to your water source.

When you arrive, take a moment to watch the water and make some observations.

When your child has observed the area and is ready to start the activity, have him or her gently lay the yarn out on the water. Make note of how it stays on top of the water using surface tension. In the event that rain is still falling, what happens when the yarn is hit by raindrops that break the surface tension?

➤ **Before you do this surface-tension string art activity you may want to dive a little deeper into getting to know your water source. Make some specific observations about plants growing in the water and how the water behaves. Here are some questions to get you started:**

Are there grasses or other plants growing up out of the water?

What does the water around them look like?

Is it moving up the sides of the plant where it breaks the surface?

How do the raindrops look when they hit the water?

Why does the water ripple when they hit?

WET WEATHER

TAKE A **DISCOVERY WALK**

CRITTER HUNT

Although it might not seem like the obvious thing to do on an afternoon that brings less than ideal outdoor conditions, a wet weather exploration can afford surprising sightings of creatures that young naturalists are sure to regard with enthusiasm. From frogs and toads to slugs and earthworms, there are plenty of local residents who are happy to get out in the rain, making it easier than usual to catch a glimpse of them.

To get the most out of your rainy-day foray, you should pack along a few things. A laminated tri-fold field guide is a great companion since it will allow your children to identify insects or animals they see, while also being waterproof. And obviously, a waterproof jacket and a comfortable pair of rubber boots will be useful as well. Boots are especially nice if your kids have an interest in wandering into areas where they might encounter mud, puddles, or both. Don't be shy about letting them! Those muddy areas near the water are perfect places to find some of the very best animals hiding.

Wherever your waterproof shoes may take you, here are a few questions you might want to pose to your young scientists along the way:

Are there areas nearby that might be more attractive to certain animals now that they are wet?

What kinds of animals might come out in the rain rather than taking cover?

How might water collecting in certain locations or on plants and trees be useful to animals?

Does the ground feel different after being rained on? Does it look different?

RAINY-DAY
LEAF PAINTING

This simple project uses familiar techniques and materials, but perhaps in an unusual way. By applying small amounts of paint to a variety of leaves gathered from neighborhood trees and plants, it becomes possible to see the intricate network of veins on each leaf and to pay particular attention to the unique patterns and outlines of individual leaves. And as paint pools in the grooves along the folds of various leaves, they become ideal materials for use in printmaking projects, making this one activity that blends art and observation in all the best ways.

WHAT YOU WILL NEED

- WATERCOLOR PAINTS
- A PALETTE OR EXTRA-LARGE LEAF FOR THE PAINTS TO REST ON
- BRUSHES OR OTHER PAINTING UTENSILS
- A TRAY OR COOKIE SHEET TO TRANSPORT PAINTED LEAVES

SCIENCE BEHIND THE SCENES: *Leaf Veins*

When you look closely at a leaf, you will notice that its surface is marked by a number of lines running in different directions. These are called "veins" and not only serve an important purpose for the leaf itself but are vital to the health of the plant or tree. A leaf's veins are generally found on the underside of the leaf, and they act as a type of "skeleton" for the leaf, giving it structure and shape. But the veins also have another very important job: they are the pathways by which nutrients and water move through the leaves, keeping them healthy and green so that they can continue to photosynthesize, ensuring the whole plant may thrive.

WHAT YOU WILL DO

Gather up the watercolor paints, palette, and brushes, and bring them outside on a rainy day.

Take a few moments to go on a leaf hunt around your yard or at your local park. For younger kids, a walk down the street in the rain for a leaf hunt can be a big added bonus. Throw in some puddle-splashing along the way! Collect leaves in a variety of sizes, shapes, and with differing textures. The larger the array of leaf options you have available for purposes of painting, the easier it is for this project to be both art activity and scientific observation.

Look around the outdoor area where you will be working and find an accessible puddle, in a safe spot, to use as a water source.

Decide which colors to use and squeeze a bit of paint onto your palette or large leaf. Choosing colors that will contrast with the leaves chosen for painting will give stunning results. Bright yellow leaves pop with purple paint, light green leaves lend themselves to a dark blue or deep red. Play with the colors you have and let your child mix paints and try different combinations, even on the same leaf!

It is easy to change the depth of color in watercolor paints by simply adjusting the amount of water used. Use more water on some leaves while employing less on others. Or try protecting some of your painted leaves from the falling drops of rain by huddling over them while you paint, while allowing others to be fully exposed to the water coming down. This will result in a variety of visual effects on the finished leaves.

Place the tray or cookie sheet in a protected spot to shield painted leaves from the rain. When you are all finished painting bring them inside to dry.

While these painted leaves are certainly beautiful and interesting in and of themselves, you might also think of pressing them onto pieces of blank paper (paint side down), using the leaves as stamps of a sort and creating permanent prints that can be hung around the house. These prints also make lovely cards to send to friends and family to honor special occasions.

Using the wide variety of leaves that you have gathered, carefully outline each leaf on a piece of paper, making sure to sketch each little point and dip in the outline of the leaf. Next, place the leaf next to the outline that you have made and observe how the larger veins can be found at particular points on the leaf. This gives the leaf structure but also plays a role in the shape, or "design," of the leaf.

In the autumn and winter, leaves that have started to decompose are often abundant. These leaves have lost their outer layer, and their delicate underlying structure can easily be seen. Take some time to hunt for and investigate these leaves. If you have a pocket microscope or magnifying glass, use it to take a closer look at your leaves. Hold them up to the light, flash light through them, or use them for this painting activity to make nature prints and other nature-inspired art.

After the painted leaves have dried, make a leaf lantern with them. Use Mod Podge, or another strong, transparent glue, to stick the leaves to the inside of a mason jar. Once they are dry, place a small candle in the jar and carefully light it. Take note of the leaf features that can be seen once the light shines through. Remember not to leave candles unattended.

MAKE A RAINBOW

A rainbow is created when white light from the sun, which has all the colors in the visual spectrum, enters drops of water. The water drops each act like a prism and bend or refract the light. Since light travels in waves, and each color has a different wavelength, each color in the visual spectrum bends at a slightly different angle. When you look at a rainbow and see the color red, you are actually seeing the drops of water that are at the correct angle from where you are to bend red light. This is true for each of the rainbow's colors. Red has the longest wavelength so it is at the "top" of the rainbow, while violet has the shortest wavelength so it is seen at the "bottom" of the rainbow. Since each person is standing in a different spot when viewing a rainbow, each person sees their very own rainbow!

You can easily make a small rainbow at home using a glass of water. Fill the glass with water, and find a place to set it that is in direct sunlight. Place a sheet of white paper next to the glass so that as the light shines on the glass, it will hit the paper too. Allow the sun to shine through the glass, adjusting the glass as needed until you see a small rainbow appear on the surface of the paper. Try out glasses in a variety of shapes as an experiment. Do they result in a variety of different rainbows?

3

LET THE SUN SHINE IN

A properly sunny day provides the greatest of motivations to head outdoors and make the most of things. The sun shining in a clear sky has an undeniable pull that leaves even the most responsible of citizens feeling conflicted about staying in to get things done versus going outside to soak up the good warm weather.

While it is certainly true that we appreciate the sun for the role it plays in our growing gardens and days at the beach, the sun does much more than many of us may realize. In looking at the types of weather that we experience in each of the seasons, one of the factors that shows up again and again is heat. When atoms inside a particular set of molecules receive new energy and begin to move faster, it is called heat. Temperature is a way of measuring heat, or how fast atoms are moving. The most important source of energy and heat for Earth is the sun. Heat from the sun is what allows life on Earth to be possible. Other planets in the Solar System that do not receive this heat from the sun are not able to have any life on them at all. Not only does heat help life on Earth to thrive, it is also one of the most important reasons why we have weather. From creating the uneven air pressure that drives global winds to providing the heat that makes the water cycle work, the sun is the superpower creating weather in our skies.

The activities and projects on the following pages have been cleverly devised to encourage you and yours to make the very most of the sunny skies, whenever you might be gifted with them. You can play with shadows and re-create the water cycle on your kitchen counter. There are opportunities to learn art techniques both traditional and abstract, as well as helpful hints for employing the sun itself as a medium in your artwork.

So, if you happen to look out your window today and see the sun shining back at you, don't hesitate to head out the back door, book in hand, ready to explore the world of art and science the heat of the sun brings your way.

SUNNY DAY SHADOW ART

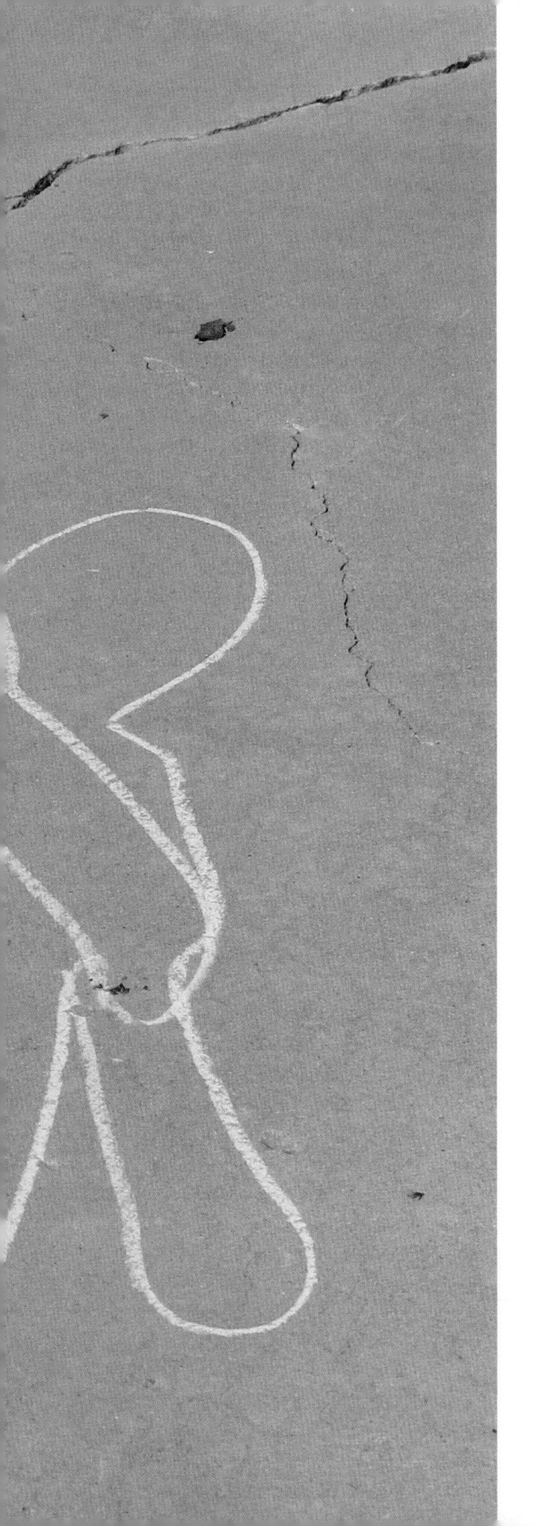

When children are very young, their observations about the position of the sun in the sky tend to be simple and straightforward. They may only go as far as noticing the difference between night and day, for example. But as children grow older and their curiosity about the way things work leads them to look deeper, they may begin to notice that the sun occupies a different space in the sky, not only as the day goes on but also as the seasons progress. This outdoor activity is an easy one, and requires few materials. Despite its humble conception, it is a good project that will give you hours of opportunity for observation, art, and, yes, some science, too.

WHAT YOU WILL NEED

- SIDEWALK CHALK
- A FLAT AREA SUCH AS A DRIVEWAY OR PATIO THAT GETS GOOD SUN EXPOSURE

SCIENCE BEHIND THE SCENES:
What Makes a Season?

You may already know that the Earth orbits, or revolves, around the sun, and that it takes one full calendar year for the Earth to complete its orbit. However, scientists believe that not long after it formed, Earth collided with another planet in space. Since that time, the Earth has sat at a slight angle on its axis. This means that at different times during the year, some parts of the Earth are tilted more toward the sun than others. The result is our seasons. The angle of the Earth in relationship to the sun means that during some times of the year, some parts of the Earth will receive more warming sunlight and have more hours of daylight, while other parts of the Earth will have less. So seasons vary around the globe. When it is summer in the Northern Hemisphere, it is winter in the Southern Hemisphere.

WHAT YOU WILL DO

Early in the morning use the chalk to outline a shadow being cast across the flat area (kids love to step in as models for this part, but anything casting a shadow will do).

Return as frequently as you like throughout the day to check on the shadow as it moves and to make more outlines as it does. Each time you visit, check the position of the sun in the sky and consider how the shadow has moved accordingly. Decide if it has moved enough to make another outline or if your child wants to wait longer.

Add to the fun by keeping a time log of each outline. Discuss how long it takes the shadow to move and why the movement occurs. Try this in different seasons and point out where it was done in summer, and if it could be done in that same spot in winter.

If you don't have a flat area (or if there is snow outside), get creative with outlining in string, pebbles, sticks, or another medium that will work in your environment.

Try playing with shadows at the beach. Place a stick upright in the sand to track the movement of the sun.

Take a series of photos at sunrise or sunset throughout the year, including one for each season, taking note of the changing position of where the sun comes up each morning or sets each evening. Place them in a nature journal or album in order to help visualize and illustrate how the Earth shifts in relationship to the sun. As a reminder to take your photos, you may want to mark your calendar for the spring and fall equinoxes and the winter and summer solstices. (Remember not to look directly at the sun. That can damage your eyes!)

Taking note of how shadows block the sun, see if your kids can feel any difference in temperature by standing in the sun and then standing in shade nearby. Use a thermometer to take the temperature in full sun and again in a shadowy area. Record the difference in your nature journal. Do this in the different seasons and note any changes in the way shadows impact temperature.

SEEING SHADOWS

TAKE A **DISCOVERY WALK**

In this activity, you can take a simple approach to make the most of a bright and beautiful day by heading out to watch shadows shift in the sun. There are no materials needed, you don't have to pack along tools of any sort, and best of all, the whole experience is provided to you by Mother Nature free of cost. Although it can be entertaining to take along a camera to snap photos of interesting shadows or a piece of chalk for making shadow art, this is by no means necessary. The shadows themselves are like nature's artwork, and just watching them flicker and dance can be entertainment enough. Here are a few fun considerations for your shadow walk:

How can you tell which shadow belongs to which object?

Is an object's shadow larger or smaller than the object itself?

Use your hand, a piece of a plant, or a stick to cast a shadow on the ground. What can you do to change the shape of the shadow? How can you change the size?

Take a shadow walk on subsequent days, at different times. How do the shadows of things in your neighborhood look different at different times of day?

What might be causing the shadows you see to appear to "dance" as you walk?

EVAPORATION
STATION
FRESH

This activity will allow you to play with evaporation at your kitchen counter or on your front porch, letting you keep a close eye on the bodies of water and what happens to them. An easy way to demonstrate the differences in evaporation of salt and fresh water, this is simple science that not only teaches about the water cycle but encourages development of observation and prediction skills as well.

WHAT YOU WILL NEED

- 2 CLEAR CUPS OR SMALL JARS
- ½ CUP FRESH WATER ***(divided into two ¼ cups)***
- 1 TEASPOON SALT ***(preferably sea salt)***
- 2 TOOTHPICKS OR STICKS
- SCRAPS OF PAPER OR WHITE STICKERS
- NOTEBOOK OR PAPER WITH PENCIL FOR TAKING NOTES

SCIENCE BEHIND THE SCENES:

"What Does Air Taste Like?"

Most evaporation comes from water in the ocean, which we know is salt water. So why don't we taste salt in the air if water vapor from the ocean is all around us?

When water molecules break free from the ocean water to become water vapor they leave the salt molecules behind in the ocean. Salt cannot become vapor at low temperatures, so the air does not become salty.

WHAT YOU WILL DO

Use the scraps of paper to make two labels: one for the salt water, one for the fresh water. Attach the labels to the toothpicks or sticks with tape or glue.

Dissolve the salt in ¼ cup of water. Fill one jar with ¼ cup fresh water, the other with the ¼ cup of salt water. Place the labeled sticks in the jars, or tape them to the sides of the jars. Place the jars in a sunny location and leave undisturbed for a few days. A covered porch works well. If they are out in the open, be sure to bring them in if rain is in the forecast, and return them to their sunny spot once it has passed.

A few things for young scientists to consider:

- WHAT DO YOU THINK WILL HAPPEN NEXT?
- WHICH ONE WILL EVAPORATE FASTER?
- WHERE IS THE WATER GOING?
- WHAT DO YOU THINK WILL HAPPEN WITH THE SALT?

Check daily to see how the water level in the jars has changed. Make note of any changes you see. When the water levels start to get really low, begin looking for any other changes on the sides of the jars (such as salt accumulation).

The rate of evaporation will vary depending on the level of humidity in your area while you are doing the activity. Look up the humidity on a weather app or the Internet, and track how it is affecting the evaporation rate of the water. You may even consider seeing if your child wants to conduct this activity during different seasons to see how the rate of evaporation changes throughout the year.

MORE TO EXPLORE

Take this science to the next level by making a set of control jars to place in the refrigerator, and also making a set of covered jars, along with the open jars from the experiment above. Follow all of the same procedures and investigate how the humidity in the refrigerator is controlled and how it will impact the rate of evaporation. Compare the three sets of jars using the measurements taken with a ruler. Let your child explore and brainstorm other ways to measure and compare the three sets.

If your child wants to take measurements to track changes in the water level more closely, simply dip a ruler into the jars to measure the water level in each one (be sure to clean the salty water off the ruler before dipping it in the fresh water). You can also use a marker to make marks on the side of the jar (if it has straight sides) and measure from the base to the mark.

This concept can be extended and played with in a number of other ways. You may want to try using jars with different heights, shapes, and so on. You can also add food coloring to the jars to see if certain colors disappear faster than others.

WONDERING ABOUT OUR WORLD

VIRGA

Although we tend to focus on water that has already reached the ground when we think about evaporation, it is actually possible for rain or snow to evaporate before it even touches down on Earth. Precipitation that evaporates before making landfall is called "virga." Virga is often the result of a storm moving through an area that has low humidity and very hot temperatures, such as a desert. But it can certainly happen in other locales as well, and it is surprisingly common! Observing virga is a great way to connect rain with the science of evaporation.

SEE WHAT TRANSPIRES

Plants don't sweat the way that people do, so transpiration is tricky to watch. But with a plant and a plastic bag you can do a fun activity that will let you see how it works. This simple, classic science activity allows transpiration to become visible to the naked eye, letting you get a sneak peek at a vital process that is happening all around you, even when you don't see it.

WHAT YOU WILL NEED

- A LEAFY HOUSEPLANT
- A PLASTIC BAG
- A PIECE OF STRING OR TAPE
- NOTEBOOK OR PAPER WITH PENCIL FOR TAKING NOTES
- A SUNNY PLACE TO PUT YOUR PLANT

SCIENCE BEHIND THE SCENES: *Transpiration*

Transpiration, or the evaporation of water from plant leaves in the environment, is a very important part of the water cycle. The movement of water from transpiration happens when water vapor is released into the air from plant leaves. Transpiration is a lot harder to witness firsthand than other phenomena in the water cycle like a rainstorm, a snowdrift, or even the steam rising from a cup of hot liquid. Indeed, the process of transpiration is very nearly invisible, unless you know how to look for it! Since some kids may have a hard time visualizing just how water can move out of leaves, it might help to explain to them that the root "trans" means movement, or to cross over. Just as trucks and trains are types of transportation moving goods and people from one place to another, transpiration moves water into the atmosphere.

WHAT YOU WILL DO

Water your houseplant with plenty of water to start off. You want to make sure it has lots of water running through its leaves for great results.

Take a moment to investigate the plant's leaves. Look on the top and underside, asking your child if they see any spots where water could come out of the leaves. Point out the veins and how they structure the leaf and provide channels for water and nutrients to flow to each leaf.

Cover a section of the leafy plant with a plastic bag and seal it tightly with tape at the stem, being careful not to break the stem or leaves.

Place the plant in a sunny location and continue to make sure that it has an ample supply of water.

Check it hourly until you see water droplets forming on the bag. Once water droplets begin to form, take some time to look at the leaves that are not covered. Ask your child to feel the leaves that are uncovered to see if the plant feels wet. Point out that both sets of leaves are releasing water but you can only see it because the bag is covering one set and the water is condensing on the bag.

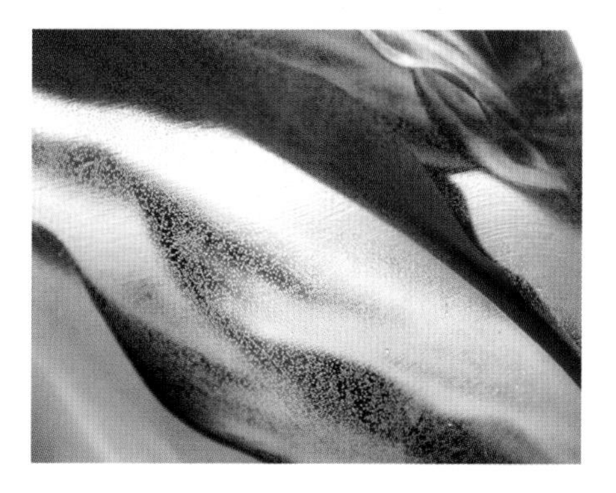

Keep checking over the next few days to make observations and note any changes. Once you have made your observations, remove the plastic bag.

The results of this activity depend greatly on the concept of condensation. What is condensation? Condensation is simply the conversion of water vapor to liquid water. This often happens when water vapor comes in contact with a cool surface, such as a glass of ice water. This is also why a glass of water is more likely to develop water droplets on the outside on a hot, humid summer day than on a cool winter day when the humidity is low.

To play around with the concept of condensation a bit more, use multiple glasses of water that are different temperatures to see if there are any observable differences in condensation. This can be done by having a glass of water at room temperature, then multiple glasses with one, two, three, or more ice cubes in each one. Once the observations have been made, leave the glasses out until all of the water droplets have gone. What happened? Where did the water go? Explain that the water has evaporated to become water vapor again, just as the water being released from the plants becomes water vapor.

THE LONGEST AND SHORTEST DAYS

Children generally know that the four main seasons in a year are winter, spring, summer, and fall (or autumn). They may not know, however, that which season we experience depends on where we live on Earth, and the seasons change each year on dates that we are able to predict. So, beyond what we can observe, how do we know when the season will change from one to another? When the celestial equator is as far away from the sun as it can be, we call that time a solstice. This happens because of the way that the Earth is tilted on its axis. There is a winter solstice, when the season changes from fall to winter, and this happens on December 21 or 22 of each year. There is also a summer solstice, and this takes place on June 20 or 21 of each year. But during the course of these seasons, the Earth continues to orbit the sun. Eventually, the Earth reaches a point in orbit where the tilt of its axis makes it in line with the position of the sun; it doesn't appear to be tilted at all. When this happens, it is called an equinox. There are two equinoxes each year: the one that begins the season of fall, and the one that begins the season of spring. The fall equinox, or the first day of fall, is on September 22 or 23. The spring equinox is on March 20.

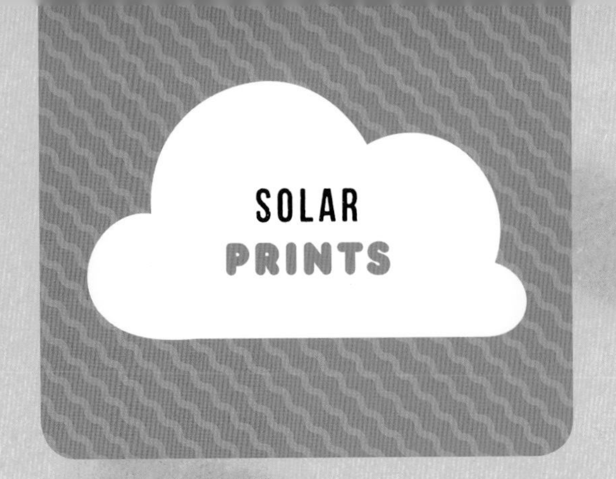

SOLAR PRINTS

Although you can certainly pop into your local craft shop and grab a package of photosensitive paper for purposes of making nearly instant sun prints, there is another way to harness the energy of the sun as an art medium. The results may require a bit more patience, but extending the sun-print process here allows for deeper exploration of the science behind the pretty prints that result. This is also a great project for taking on the road, since the materials needed are so simple and so few, and every new environment you visit will present unique opportunities for collecting and arranging specimens for stunning sun prints.

WHAT YOU WILL NEED

- BRIGHTLY COLORED CONSTRUCTION PAPER ***(lower-quality papers will fade faster)***
- NATURAL ARTIFACTS YOU'VE GATHERED SUCH AS LEAVES, STONES, BRANCHES, AND FLOWERS
- A FLAT SURFACE WITH GOOD SUN EXPOSURE
- TAPE ***(optional for windy days)***

SCIENCE BEHIND THE SCENES: ***Bleached by the Sun***

Most of us recognize that sitting out in the hot sun for a good spell will lighten or bleach favorite fabrics and other things with a distinct color to them. But the science behind this phenomenon is perhaps less well known, even though it is driven by the same power that makes so many other things happen: the sun and its powerful energy. In the case of things with color, the sun's energy changes the chemical structure of the molecules that make up pigments. As the molecules break down and change, so does the color of the item in question. "Photodegradation" is a word used to describe color fading. So, as the sun rains energy down upon the everyday objects in our world, vibrant greens fade to muted yellows and bright reds become softer shades of pink or orange.

WHAT YOU WILL DO

Take a walk in the woods or browse the backyard and gather up a good armful of things to use for your print.

Lay the construction paper out on a flat surface and spend some time arranging your objects on the paper to your liking. If there is a slight breeze, tape the paper down and choose objects that will not easily be blown away.

Leave your solar print setup sitting out in the sun for a number of hours. Check periodically to see how the pigment around the objects begins to fade or change color while the areas underneath the objects maintain the original hue and brightness of the paper. Once the color has faded enough to see a print, remove all of the objects and enjoy your print.

These prints make great cards for special occasions. Simply fold the construction paper in half (or cut off a section to fold) and paste a bit of plain paper on the inside to make a space to write or draw a special greeting for the intended recipient.

By using construction paper instead of photosensitive paper you can slow the process down and help your child to make the connection that it is the sun that is fading the chemicals and dyes in the paper, therefore creating the design.

You can also try to cover paper with different brightly colored paints and do the activity with your painted papers. Painting a rainbow and covering only parts of the rainbow with sticks or other straight objects would work well for this type of investigation. Do some shades retain their color better under the sun than others?

SAND PAINTING

This is a great project for families on the go. Other than a bottle of school glue, it can be done using only materials gathered from whatever beach or backyard you happen to be visiting that day, and because it is a drawing-based activity, it is endlessly flexible and can be done again and again with a different result each time. So, the next time you are headed for the shore, or even to a local sandbox, think about tossing a bit of glue into your backpack. You'll be glad that you did.

WHAT YOU WILL NEED

- DRY SAND ***(at the beach or from the sandbox)***
- WHITE NONTOXIC SCHOOL GLUE
- ROCKS OR DRIFTWOOD

SCIENCE BEHIND THE SCENES: *Erosion at Work*

As we explore the natural world, many of us might think here and there about what things used to be. Butterflies that used to be caterpillars, for example. Or baby birds that used to be confined to a colorful egg. And then there is sand. It might be easy for kids to assume that the sand beneath their toes at the beach or on a local riverbed was always there, but geologists would beg to differ. Actually, sand is the result of the Earth's weather. Over time, pieces of rock are exposed to wind and water, or they become embedded in glaciers that move them along the Earth's surface. All of these weather-related actions cause the rocks to be broken down over time, turning larger pieces of rock into the small stuff we call sand.

WHAT YOU WILL DO

Find a rock or a piece of driftwood with a somewhat flat surface. Choosing a rock or wood that contrasts with the color of the sand will result in a drawing that stands out.

Using the glue, create a shape, small scene, or abstract dot-and-line drawing on the flat surface of the rock or piece of wood. Squeeze glue directly from the bottle so that you get a fine line.

Gently sprinkle the glue with the dry sand until all of it is covered. Turn the rock or wood over to let any extra sand fall away, revealing the sand drawing.

Let your masterpiece dry in a sunny spot.

These drawings make great paperweights or nature table decorations, but they can also be left behind as an unexpected treasure for another beachgoer to find.

You can keep making art from your sand paintings long after you have finished them. After the paintings are fully dry, cover them with plain paper and use a crayon to make a rubbing from the sand art. Just make sure to be a bit gentle so as not to rub the sand off your painting!

Painting with sand also presents a good opportunity to think about where sand comes from in the first place. Sand is made from the breakdown of larger rocks over a period of time, making rock exploration a perfect extension activity for sand painting. Gather a handful of rocks with different colors and textures. Rub the rocks on hard surfaces, such as a driveway or sidewalk, and try crushing them with other rocks as well. (Please use eye protection when necessary.) Which rocks crumble, flake, or begin to otherwise fall apart under pressure? Which ones leave colorful marks behind when they have been rubbed against something?

Rock erosion plays an important role in creating sand, but it is also a somewhat abstract concept for younger naturalists. This simple activity is a good one for helping children to understand a bit about how rain and water flow play a role in erosion, and what the effects might be. Fill a cookie sheet with small rocks or gravel, and on top of that add some sand, and using a funnel direct the rain to a specific area on the cookie sheet to simulate a flash flood or the beginning of a river. The sand should wash away, leaving the gravel (riverbed) behind. Rain could also be collected in a jar then poured into the funnel to make this faster for small children, or even redirected from a rain gutter on your house.

NANCY HOLT

Although she was born and raised on the populous East Coast of the United States, Nancy Holt eventually became known for the land art exhibitions she created in the vast, empty landscapes of the desert Southwest. An important member of a pioneering group of artists who incorporated elements of the environment into their artwork, Holt emphasized connection with nature in the large-scale sculptures that she created, encouraging people to shift their perspective about the natural world and their place in it. One of Holt's more well-known works are the "sun tunnels," four gigantic concrete tubes placed in the Utah desert where the effects that they create vary based on the movement of the sun across the sky and through the seasons. Each tunnel is also outfitted with a collection of holes, strategically sized and placed to mimic a selection of constellations. Holt sought to integrate the earth and sky through the tunnels, and through her work in general.

Nancy Holt was a lesser-known artist than some of the others looked at in this book. However, a bit of searching online or at your local library will reveal that she was part of many collaborations with other artists who used the environment as a setting for their artistic endeavors. Photographs of some of these exhibitions can be found on the Internet, but learning more about Holt's work might also be the perfect reason to introduce your budding researcher to the local reference librarian as well.

WATERCOLOR CUBE PAINTING

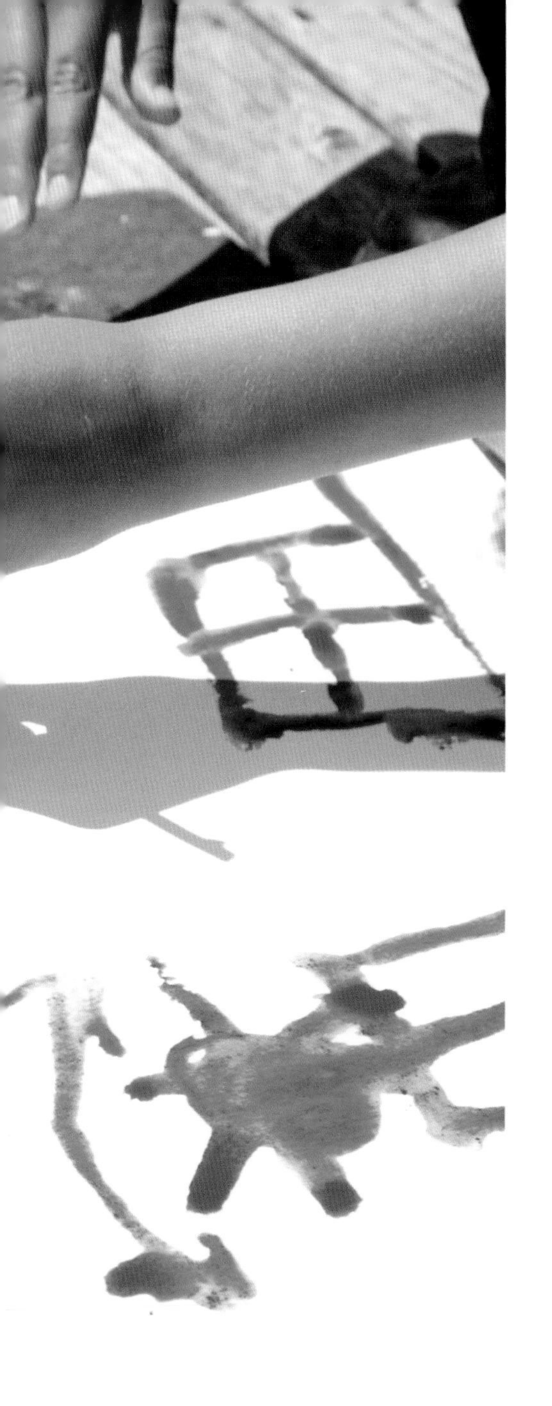

These watercolor cubes make for a fun twist on a classic science experiment. By making ice cubes from watercolor paint, you and yours can engage in some entertaining outdoor art-making, while watching the principles of heat absorption in action at the same time. A built-in handle makes the cubes easy to paint with, and while you are creating your latest refrigerator hanging, you can dive into a discussion of the melting rates of the different-colored cubes and just why that might be happening.

WHAT YOU WILL NEED

- TUBE WATERCOLOR PAINTS
- AN ICE CUBE TRAY
- STICKS
- A PLACE TO FREEZE YOUR CUBES ***(in the freezer, or outside on a winter day)***
- WATERCOLOR PAPER

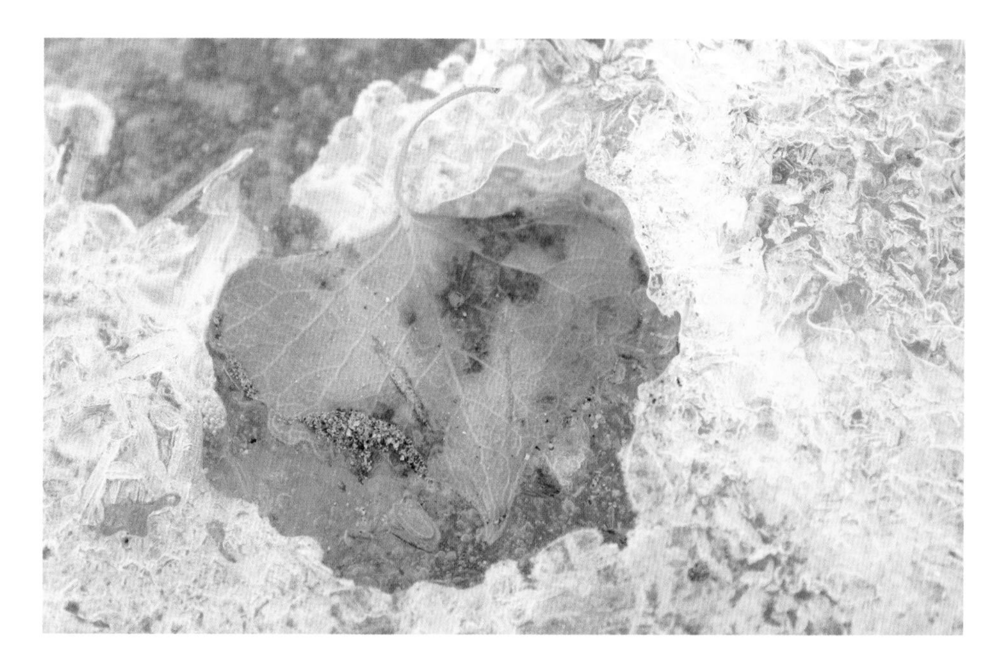

SCIENCE BEHIND THE SCENES: *Hot in Here*

The presence of heat is simply part of life on Earth. Like the rest of us, kids certainly know when a slide is hot or their ice cream is melting quickly in the summer heat. But they probably don't know how heat actually works. Heat is created by the movement of atoms in our world, and since everything is made of atoms, there is plenty of movement happening at any given time. And yet, some objects absorb heat more readily than others. You may notice that dark-colored things seem to heat up more quickly on a sunny day. This is because darker colors absorb heat more effectively than lighter colors do. As a result, darker-colored objects heat up more quickly than light ones, making them more likely to melt or be hot to the touch than lighter-colored objects nearby.

WHAT YOU WILL DO

Fill the ice cube tray with water as if you are making ice cubes. Put a small dollop of paint into each section. For younger children it might be nice to stick to two or three complementary colors.

Using a stick or paintbrush, mix the colors until well blended with the water. As the paint freezes some will naturally settle in the bottom of the tray. Not to worry—this actually makes for a nice splash of extra color here and there once painting begins.

Place the tray in the freezer for about an hour, checking on the freezing process. After the top of each cube has frozen over, insert the sticks into the cubes. Use other items in the freezer to help the sticks stay upright if needed. You could also cover the tray with foil and carefully insert the sticks through it to help them stand up.

Once the cubes are frozen, carefully loosen them from the tray. If necessary, run some warm water on the bottom of the tray to help release the cubes, being careful not to get water in the top of the tray as it could mix your colors.

Keep the cubes organized in the tray and take them outside. Set up a larger piece of watercolor paper to paint on. Use the watercolor cubes to paint: the bottoms of the cubes can be used to paint in broad strokes; the edges are particularly good for fine detail. After you've started painting, you might want to dip your paint cube back into the tray to wet it and gather more paint that has stayed on the bottom of the tray.

During the painting process you may want to take the opportunity to talk about which colors are darker and which ones seem to be melting faster. The melting rates will depend on how hot it is outside and the heat absorption of the different colors.

This project is a great chance to explore color mixing. Try creating new colors by combining the primary ones you have available.

If you have a hill in your yard or in a nearby park, consider using a larger sheet of paper and positioning it so that the paper is oriented downhill. Set out the watercolor cubes on the paper so that as they melt, they will also move downhill on the paper, creating interesting patterns and allowing you to observe how faster melting rates also result in faster movement down the hill!

Once the painting is finished, consider setting the leftover cubes out on a sheet of watercolor paper to observe how they melt and create works of art all on their own.

COLORFUL SKIES

What makes the sky so colorful at sunrise and sunset? Light is divided into wavelengths that we call the visible spectrum. These are the same wavelengths that we can easily observe when looking at a rainbow. The bright, beautiful colors sometimes observed at sunrise and sunset are caused by a phenomenon called scattering, which is also what makes the sky look blue at midday. Scattering redirects light as it passes through gases, water vapor, and other particles in the air. The short wavelengths of light, such as blue and violet, are scattered more readily than other wavelengths, so our eyes see blue sky. As the sun is rising or setting, the light must travel through more of the atmosphere and more particles are available to redirect light, including the easily scattered blue and violet, leaving the other longer wavelengths of yellow, orange, and red to reach our eyes.

This phenomenon is also related to the shift of light during the seasonal transitions. People often talk about the golden glow of autumn light. This glow is directly related to the position of the Earth on its axis and the angle at which the sun's light is passing through the atmosphere during different seasons. The high, strong sun shining down in summer provides a bright clear light that begins to shift after summer solstice toward the more golden glow of autumn and darker gray of winter, and then shifts again in spring and starts the cycle once more.

DOTS IN THE DIRT

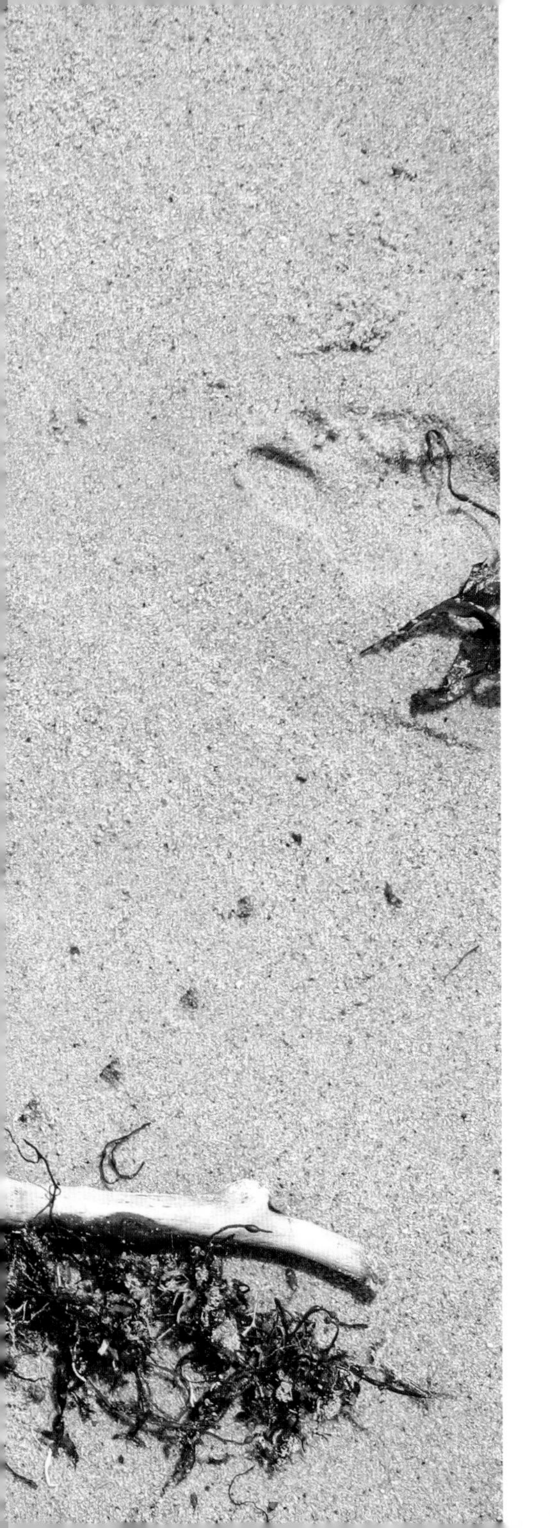

"Pointillism" is the term used by artists to describe a painting technique where dots of color are used to create the image the viewer sees on the canvas. For our purposes here, we are expanding the definition of pointillism just a little, adapting the method to make it a perfect vehicle for making outdoor "paintings" in your favorite sandy spot. From beach days to backyard explorations, this activity is a great one for incorporating an extra bit of art in your play.

WHAT YOU WILL NEED

- AN AREA WITH DAMP SAND OR DIRT
- OTHER NATURE FINDS THAT MAY BE USED TO DECORATE THE "DRAWING," INCLUDING TWIGS, LEAVES, SHELLS, OR SEAWEED ***(great for making flags with a stick)***

SCIENCE BEHIND THE SCENES: ***Shadow Science***

Shadows are a familiar occurrence that even the smallest children find fascinating. And although the casting of shadows on a sunny day can seem slightly mysterious, the science behind shadows is fairly simple. Basically, shadows appear when the sun's light is blocked by an object between the sun and the Earth. The lower the sun is in the sky, the longer the shadow it casts. Astute observers may have already noticed shadows on Earth have just a bit of color to them. Because of the way that different colors reflect light, outdoor shadows on our blue-skied planet appear slightly blue as well.

WHAT YOU WILL DO

Using a finger or stick, poke holes in the sand or dirt, creating a picture with the dots. Be sure to poke the sticks down deep enough that the resulting hole casts a shadow inside that can be clearly seen when looking at the drawing from different angles.

Use bits of nature to decorate the picture if you like. Think about adding some flags to the perimeter of your picture, constructing them out of twigs, seaweed, or leaves. Features such as chimneys can be accentuated with "smoke" made from leaves or seaweed, and shells can act as fruit hanging from a tree. You might even lay down driftwood or sticks as rows in a garden. Note how the inclusion of different objects, and the shadows they cast, alters the piece of art by drawing attention to some areas.

Sun casting shadows on the dots

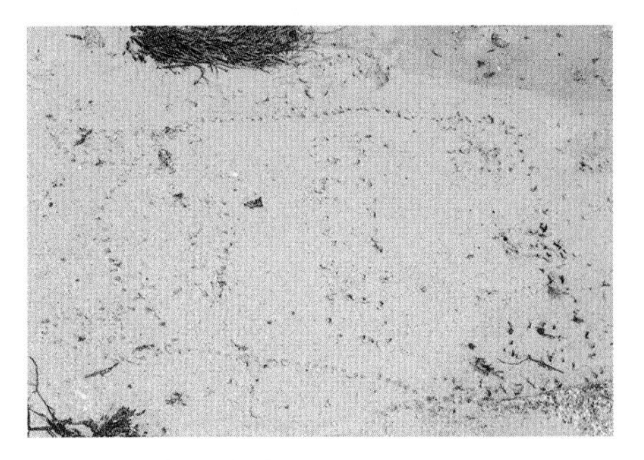

The dots without shadows

As the day progresses and the sun moves across the sky, observe the picture to see how the shadows within the dots (and decorations) move and make the picture look slightly different. You can also move from one side to the other to see how the picture changes.

Take a series of photographs of your picture as the shadows it casts change across the day. Assemble the pictures sequentially and display them.

You can also use this same technique to create a painting or a piece of sculpture that can be kept and displayed for a longer period of time. For a painting, simply use the dot-making pointillism technique to create images with paint and paper that are made up of dots of different sizes. For a piece of art made from clay, try rolling out a slab of kid-safe clay and using a stick to create an image made of dots on the surface. Allow it to dry, and then put it on display in a sunny spot to watch how it changes throughout the day.

TRACK THE TIDES

Tides are an amazing peek at the relationship Earth has with the moon and sun, along with the roles gravity and inertia play in our world. While the whole business of tides is quite complex, there are ways to break the process down into simpler terms. The sun plays a minor role in the movement of the oceans to create the tides due to its gravitational pull on the Earth. But the moon is really the star of the show. The gravitational pull of the moon, along with the inertial force caused by the rotation of the Earth, causes the high and low tides that oceans experience approximately every six hours.

This rise and fall of the tides is caused by bulges in the seawater that are created on both the side of the Earth closest to the moon and the side farthest away from the moon. When Earth is aligned with the sun and moon (as it is at the new and full moons) these forces are strongest and create the greatest variation in the tides. We call this the spring tides. The name is perhaps misleading, as these tides are not seasonal at all but instead are bimonthly occurrences that follow the phases of the moon. If you live in a coastal location (or have access to a coastal webcam to watch the tides on your computer) it is fun to observe the tides during the new and full moons to see the greatest variation in the tides on a daily basis.

On a day at the beach, track the retreating tide by placing a line of rocks at the high-tide mark. Then as the tide goes out, place another line of rocks at the new tide mark. Continue doing this over the course of your beach day to track the retreating tide.

You may want to make some seaweed nature flags to define your project area for others. Think about snapping a photo of your tide markers when the tide is at its lowest point. It makes for a lovely bit of refrigerator decoration or inspiration for a more involved art project such as rendering the scene in paint or pencil.

4

WHERE THE WIND BLOWS

When you walk out your front door and feel the wind blowing through your yard, it is natural to think about the wind as air that moves. You can see the trees sway as the wind blows and feel the breeze in your hair. But the wind that moves through your own yard is part of a more complex and global wind pattern that affects the Earth as a whole.

Wind is the movement of air from one place to another. As the Earth is heated by the sun's rays, some areas of the planet warm more quickly than others. In areas with warmer air, the air will rise. When it does, cooler air moves in to replace the warm air that has moved away, creating wind. This wind, in turn, creates much of what we think of as "weather." And it often captures the popular imagination at the same time.

When we think of wind, we often think of hot air balloons and kites floating across the sky, of blustery afternoons spent indoors cozied up under a blanket. We think of dandelion seeds lifted by a gentle breeze or of trees bent by the force of fierce gusts that bring along torrents of rain. The wind, despite not being visible in and of itself, is the source of much of what we see in the natural world. It creates movement of things from one place to another, and this movement, in turn, keeps the world running as it should.

Throughout the pages of this chapter, you will have an opportunity to explore the nature of wind in the world. You will learn some ways that wind can be felt, harnessed, and even observed, despite the fact that it isn't quite as easy to see in action as a snowstorm or the rays of the sun. The activities here capture the power as well as the whimsy of the wind, and the way that it affects our environment. And, you will also find a couple of experiments that get at the heart of wind science, yielding surprising results along the way.

This chapter is one to dive into with everything that your creative and curious spirit has to give. Fold, construct, experiment, and observe. Watch as wind makes things happen, and also as it encourages them to come apart.

Feel the wind on your face, and head straight into it.

LEAF-RUBBING PINWHEELS

Unlike the sun or snow, the wind isn't something that we can see. We see flags flap or trees sway because of the action of the wind, but we aren't actually seeing the moving air itself. Scientists solve this conundrum with the use of special tools and instruments designed for measuring the wind. For our purposes here, some pretty paper pinwheels will do just fine. Colorful pieces of paper outfitted with leaf rubbings for extra pizazz easily transform into basic pinwheels that show the movement of the wind and promise to be as fun to play with outside as they are in.

WHAT YOU WILL NEED

- A SQUARE PIECE OF PAPER APPROXIMATELY 6" × 6" ***(Origami paper with one white side works well if it will not be used in very strong winds. Strong winds will require something thicker, such as card stock.)***
- A PUSHPIN
- A SMOOTH STICK OR PENCIL WITH AN ERASER
- CRAYONS
- FALLEN LEAVES FOR RUBBING
- RULER AND PENCIL
- SCISSORS

SCIENCE BEHIND THE SCENES:

Wind around the World

When kids think about the wind, they generally consider the current conditions outside their own home or school. That is, after all, what they can feel when they step out the door. So when contemplating the wind, it can be good fun to think globally instead. To this end, consider the Coriolis effect. The Coriolis effect is how scientists describe the way that the Earth's rotation affects the wind. The spinning of the Earth on its axis forces winds in the Northern Hemisphere to bend to the right, while winds in the Southern Hemisphere bend toward the left.

WHAT YOU WILL DO

On the white side of the paper, draw two lines from corner to corner making an X.

Place a leaf under the paper, white side up, and rub with the crayons until the leaf outline shows up on the paper. Change the position of the leaf so the tips line up with the X and the design shows up when the pinwheel is folded down. Try different leaves and colors, or possibly even a different leaf for each tip of the wheel.

Cut approximately halfway down each of the four lines toward the center of the X. Fold down one flap from each corner to meet in the middle. You can poke the pushpin through each point as you fold them down to make it easier to collect them together. (Taping as you go will also work for this.)

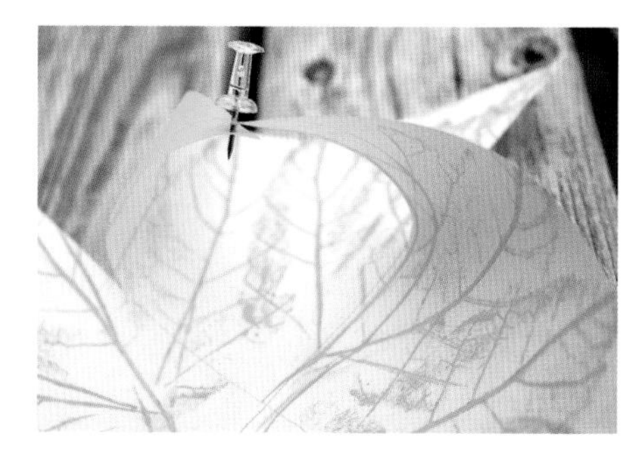

Once all of the corner points are in the center, poke the pin through the middle of the X and then into the stick or pencil eraser, leaving some space for the pinwheel to spin. If using a stick, whittle one end to be flat so that the wheel does not get hung up on the bark or bits poking out from the twig. This is a great opportunity for an older child to practice their skills with a pocketknife.

Hold your pinwheel up to catch the wind, angling it in different directions as needed to get the most movement.

Go to different locations around your yard or down your city street to check for wind direction by holding the pinwheel different ways to catch the wind.

Make a collection of pinwheels of different sizes and using paper of various thicknesses. Keep this collection by the door ready for windy-day play. As they blow in the wind take note of which pinwheels handle different types of wind better and spin more consistently in the breeze.

Try to find places where the wind blows with more or less force. Do buildings affect the wind direction? Does the pinwheel stop working when it is turned the opposite direction? Do hedges or trees affect the strength of the wind? Check often and keep a chart making note of wind direction.

Does the movement of the pinwheel change with the season, time of day, or when a storm is blowing in?

For the exceptionally interested young meteorologist, this is a good time to break out the compass to fine-tune those notes and make more precise documentation of wind direction for future reference.

AURORA BOREALIS

Aurora borealis, often called the "the northern lights," is the term used to describe sheets of colored light that can appear across the sky when conditions are just right. Aurora was the Roman goddess of dawn (or sunrise), and "borealis" is the Greek name for the north wind. So aurora borealis means "the dawn of the north wind." There are many types of auroral displays to be seen in the night sky. Sometimes the light blinks, other times it looks as if it is still, and at times it looks as if it is a fast-moving wave of light. All of these displays are caused when solar radiation from solar winds strikes particles in the atmosphere. This interaction makes those particles glow. This phenomenon also happens in the Southern Hemisphere, where it is called the aurora australis. "Australis" is Latin for "south."

To bring the idea of the auroral displays down to earth for your family, spend some time looking at photos of these stunning displays in books or on the Internet. Or if you are lucky enough to live in a place where you can observe the phenomenon firsthand, take the kids outside to make some observations of the amazing spectacle. As you look at the northern lights, take a mental photograph of the colors and movement of the lights. Then get out a set of vivid watercolors or chalk pastels and create a set of artistic renditions of your favorite "photos." If drawing with chalk pastels, explore using black paper as a background to re-create the stark contrast of those vibrant colors in the night sky. Hung on the living room wall or on a child's ceiling, they will be a colorful reminder of how very wild and beautiful the world of the night sky is.

ATMOSPHERE
MODEL

Think of this fun and funky project as a terrarium with a twist. Instead of being outfitted with a collection of pretty plants, this particular model is decorated with handmade miniatures depicting the Solar System hovering beyond the outermost layer of the atmosphere. This activity gives an opportunity for kids to explore the somewhat abstract concept of the Earth's atmosphere in a more concrete manner. By simulating the position of the various layers of the atmosphere inside an empty jar or vase, kids will have a hands-on method for considering the importance of the invisible things in the world around them.

Exosphere
Thermosphere
Stratosphere
Mesosphere
Troposphere

WHAT YOU WILL NEED

- LARGE GLASS CONTAINER OR VASE
- ABOUT ONE TO THREE CUPS OF ROCKS OR GLASS PEBBLES FOR EACH COLOR. WE USED FIVE DIFFERENT COLORS, ONE TO REPRESENT EACH LAYER OF THE ATMOSPHERE. THE AMOUNT OF PEBBLES NEEDED WILL DEPEND ON THE SIZE OF YOUR VESSEL. THE LARGER THE CONTAINER, THE MORE YOU WILL NEED TO CREATE LAYERS THICK ENOUGH TO STAND OUT. PEBBLES CAN BE GATHERED FROM YOUR YARD OR PURCHASED AT A CRAFT SHOP. IF YOU DO NOT HAVE A WIDE VARIETY OF COLORS AVAILABLE, YOU MIGHT CHOOSE TO PAINT THE ROCKS VARIOUS COLORS INSTEAD.
- WOODEN DOLL HEADS IN A WIDE RANGE OF SIZES TO DEPICT THE SUN AND PLANETS, OR MODELING CLAY ***(gray or various colors for each planet).*** YOU MAY WANT TO HAVE A DISCUSSION HERE ABOUT WHETHER OR NOT TO INCLUDE PLUTO AND SOME OF THE OTHER DWARF PLANETS. WE INCLUDED PLUTO IN OUR MODEL BECAUSE WE HAVE A SOFT SPOT FOR THE DEMOTED PLANET. CHILDREN MIGHT ALSO WANT TO INCLUDE STARS OR THE ASTEROID BELT. IN THIS CASE, THE SKY IS NOT THE LIMIT!
- WIRE HANGERS OR SIMILAR STURDY-GAUGE WIRE
- WIRECUTTERS
- WATERCOLOR OR ACRYLIC PAINT
- PAPER TO PRINT LABELS OR STICKY LABELS
- GLUE

SCIENCE BEHIND THE SCENES:
What's in an Atmosphere?

An atmosphere is the layer of gases that surrounds a planet. Within the layers of Earth's atmosphere are mechanisms that protect us from the heat of the sun, shield us from objects falling to Earth from space, provide the gases we need to breathe, and create our weather.

The layers of the atmosphere are as follows:

- THE EXOSPHERE IS THE OUTERMOST LAYER OF THE ATMOSPHERE.
- THE THERMOSPHERE IS KNOWN AS THE UPPER ATMOSPHERE.
- TOGETHER, THE STRATOSPHERE AND MESOSPHERE MAKE UP THE MIDDLE ATMOSPHERE.
- THE TROPOSPHERE IS KNOWN AS THE LOWER ATMOSPHERE. THIS IS ALSO THE LAYER OF THE ATMOSPHERE WHERE WE LIVE.

WHAT YOU WILL DO

Have your child layer the rocks or glass pebbles carefully in the vase according to the layers of the atmosphere as follows: troposphere (on the bottom), stratosphere, mesosphere, thermosphere, exosphere (at the top). You may want to use tan, blue, green, or another light color for the troposphere (the layer closest to Earth) and a darker color for the exosphere (which extends to outer space).

Label each layer of the atmosphere on the outside of the glass container using the printed labels and a glue stick or the sticky labels.

Next, paint the wooden doll heads, or make planets out of clay (poking a hole in the bottom of each one to accommodate the wire) and paint them after they have dried. We used watercolors to paint the wooden planets in washed, muted tones. This is a great time to pull out the space books that refer to planet size and ratios, along with guidance in color choices. Ask your child how much detail they would like to add. Some children enjoy the simplicity of a monotone collection of planets, while others want more details such as rings, clouds, and storms.

Cut the wire to appropriate lengths, depending on the size of your container and how high you would like your planets to hover above your atmosphere. Insert a piece of wire into the bottom of the sun and each planet, making sure that it is long enough to allow the planet to be firmly stuck into the pebbles in the container. Glue as needed.

Place the planets into the rocks above the exosphere, arranging them however you like. They can dry in place above the atmosphere. Step back and enjoy the grand view of our atmosphere and Solar System!

Extend this project by researching the temperature of the different layers of the atmosphere and choosing (or painting) rock colors that correlate with those temperatures: red for the extremely hot thermosphere, blue for the very cold mesosphere, and so on.

Ask your child to draw small pictures of things that might be found in various layers of the atmosphere on white labeling stickers. Cut the drawings out and stick them on the appropriate layer. Some things to research and draw include the orbits of the space shuttle and human-made satellites, the layer where meteors burn up, where the aurora borealis occurs, where weather (such as clouds and rainbows) occur, and where birds fly.

IS IT A METEOR OR A METEORITE?

Only a small percentage of meteors falling into Earth's atmosphere actually make it to the Earth's surface. The ones that do are called meteorites. Here is a fun poem to help you remember the difference between a meteor and a meteorite!

Meteors fly across the sky
And make it shine so bright.
The one that makes it to the Earth
Is called a meteorite.

—Fionna

To do some kitchen-counter science around the idea of meteorites, try employing this simple method of simulating what happens when a meteorite makes contact with Earth: First, layer flour and cocoa powder in a small baking dish. Then, drop small rocks or marbles into the dish at different angles to approximate what happens when a meteorite hits the Earth's surface and an impact crater is formed. Observe how the shapes of the craters change depending on the direction and size of the projectile. Look at the debris that scatters away from the impact zone. If your child is enthusiastic and wants to try again, simply shake the pan a bit to flatten the flour and dust with another layer of cocoa powder.

AIR PRESSURE
EGGS

This eggs-periment is one that has been a favorite of school science teachers for decades. And for good reason: the "wow" factor is undeniable, and on top of that, it teaches some great fundamental weather science. The materials are minimal, but the effects of this easy experiment are anything but ordinary. Even the adults in the house will be gathering around the table to watch as the egg appears to be sucked down inside the bottle by nothing more than the air. Which is exactly what is happening.

WHAT YOU WILL NEED

- A PEELED HARD-BOILED EGG
- A GLASS BOTTLE WITH AN OPENING SLIGHTLY SMALLER THAN THE EGG
- HOT WATER
- A BOWL OF ICE WATER

SCIENCE BEHIND THE SCENES:

What Does Air Pressure Do?

The importance of air pressure in creating the weather conditions in our world cannot be overstated. Indeed, air pressure is what makes the weather system on our planet tick, so to speak.
From a child's perspective, it may seem strange to think about weighing something that you cannot see, but that is exactly what meteorologists do when they measure the pressure in the air around us. Air, like all matter, has weight. When air is heated, its molecules have more energy, move more quickly, and expand outward, lowering the air pressure. When air is cooled, its molecules will move slowly and tend to stay together. This will result in a rise in air pressure. Wind is created when air begins moving from an area where air pressure is high to one where the pressure is lower, and the air molecules will have more space.

WHAT YOU WILL DO

Begin by filling the bottle with hot water. Water from a tea kettle that is nearly, but not quite, boiling works well. This step is a good one for an adult to be in charge of. Set the bottle aside for about four or five minutes.

Next, empty the hot water from the bottle, and pop the egg on top of the bottle's opening. Doing this quickly will help keep heat in the bottle. Place the bottle in the bowl of cold water, and watch the egg get pulled down inside!

When the air inside the bottle is heated, it expands. When the same air is suddenly cooled, and the egg is sitting atop the bottle, the air begins to contract, sucking the egg down into the bottle as it shrinks.

To get the egg out, turn the bottle upside down, and carefully pour hot water over the bottle (hot tap water usually works fine). The egg should slide back out of the opening.

This activity can also be done using a small water balloon filled so that it is about the same size as an egg.

Scientists who study weather use a tool called a barometer to measure changes in air pressure. Although it isn't nearly as exact in its ability to measure, it is possible to make a homemade barometer and use it to get an idea of how changes in air pressure are generally determined. To do this, take a balloon and cut off the mouth (the part that you blow air through). Stretch the balloon over the top of a regular kitchen jar, using a rubber band to secure the balloon in place. Tape a drinking straw to the top of the jar, so that about a quarter to a third of the straw hangs over the edge of the jar. Set the jar on a counter or windowsill, and tape an index card to an upright surface behind the end of the straw that sticks out. Draw a line just where the straw lands in front of the paper. Check your barometer here and there in the days to come, and use a pencil to mark where the straw lands. Generally speaking, the high air pressure associated with good weather will push the balloon inward, making the straw rise. The lower pressure associated with less ideal weather will make the balloon expand, pushing the straw down.

CHRISTO AND JEANNE-CLAUDE

Christo and Jeanne-Claude is the collective name used for a husband-and-wife team of artists, known for their grand visions and unusual sculptural installations. Educated in Europe, the pair arrived in the United States in the 1970s, making a permanent home for themselves in New York City. It was in New York's Central Park that Christo and Jeanne-Claude eventually staged what is perhaps their best-known exhibition, a project many years in the making that the pair named ***The Gates***.

Set against the bleak backdrop of a New York winter, ***The Gates*** consisted of nearly twenty-three miles of glowing orange fabric flags attached to sturdy steel poles and placed along the pathways of the park. Christo and Jeanne-Claude spent more than twenty years meticulously refining their vision of what ***The Gates*** would look like when it was complete; the exhibition then lasted for a mere sixteen days. Truly unique in both concept and scale, ***The Gates*** is remembered as an example of art that used weather conditions and the natural environment to give manmade materials new meaning.

A number of books have been published about Christo and Jeanne-Claude, many offering stunning photographs of their imaginative installations. Among the best of these is ***Christo and Jeanne-Claude: Through the Gates and Beyond*** by Jan Greenberg and Sandra Jordan. The book is noteworthy for many reasons, but perhaps most of all because the intended audience for the book is actually children. The story is a testament to vision and perseverance, and young artists are sure to find it inspirational.

NATURE KITES

While this kite does not fly like a typical kite, catching the breeze to soar high into the sky, it will still provide plenty of entertainment for kids exploring the nature of wind on a breezy day. These small kites make simple work of demonstrating how the creation of a bit of wind can result in lift just strong enough to send small objects upward on the air, which kids will enjoy experimenting with. Add speed, change direction, change the length of the kite string: the possibilities for inquiry are many. And as an added benefit, the materials for the project are likely to be found in a simple search of your backyard and kitchen junk drawer, making it an easy and adaptable activity that doesn't require much fuss.

WHAT YOU WILL NEED

- WAX PAPER
- A LEAF
- TWO THIN STICKS
- AN IRON
- OLD TOWELS TO PROTECT YOUR IRON AND WORK SURFACE
- A SPOOL OF THREAD
- A LENGTH OF THIN YARN APPROXIMATELY 25 INCHES LONG
- GLUE DOTS, STRONG GLUE, OR TAPE

SCIENCE BEHIND THE SCENES:
How Do Clouds Stay Aloft?

A fluffy cloud bank shifting slowly overhead is always worth watching, and often leaves us wondering just how it is that clouds manage to stay aloft.

There is a combination of scientific factors at work when it comes to the ability of a cloud to stay up in the sky, but the most important have to do with lift and size. Lift is the force created when air moves in an upward direction, taking with it objects light enough to float. Clouds are light enough to be affected by lift because the droplets that make the body of a cloud are incredibly tiny. It takes many cloud droplets coming together for them to be heavy enough to counteract the airlift pushing them up. But when this finally happens, these droplets will indeed fall to Earth as precipitation.

WHAT YOU WILL DO

Plug in the iron to warm up, and set it on high.

Create your kite tail by cutting three pieces of yarn about an inch each and tying them to the longer length of yarn at equal distances apart.

Cut two pieces of wax paper large enough to cover your leaf with a few inches of paper extending beyond the edge of the leaf. This will vary depending on the size of the leaf. We used an average-size maple leaf for our kite.

Place the leaf between the two pieces of wax paper and place the wax paper sandwich on a towel. Place another towel over the top to protect the iron. Iron the wax paper over the leaf, making sure to seal the leaf between the two sheets.

Once the leaf is sealed and the papers are securely attached together, cut out a standard kite-shaped diamond with the leaf in the center. While you certainly do not have to break out the ruler for this part, feel free to add some

math into the mix if your child would like for their kite measurements to be more precise.

Once you have your kite cut out, attach the sticks to the underside by making a cross shape and securing each end of the sticks with a glue dot. (We found that glue dots work wonderfully and there is no wait time for drying involved.) Next, place your tail at the base of the kite using another glue dot at the bottom.

Secure the spool of thread to the kite by tying one end of the thread to the sticks where they cross.

Now find a windy spot and watch your kite fly!

You can make a variety of kites of different sizes with leaves from a number of different trees, and test out the differences in the way that they fly.

You can also make variations to the tail of the kite, adding or taking away ties or weight to see how that might affect the flight or stability of the kite. You may even want to try flying a kite without a tail at all!

Forgo the wax paper and simply attach the sticks to a large sturdy leaf. How does it fly? What are the differences between the leaf-only kite and the leaf-with-wax-paper kite?

LET'S LISTEN

TAKE A
DISCOVERY WALK

One of the things that people often mention when talking about wind is the sound that it makes. We use words like "howl," "whistle," and "rustle" to describe the noises that the wind makes as it blows. Perhaps this is because wind isn't visible in the way that other elements of weather are, so we need to use another of our five senses to be aware of it. But it could also be because the wind really does make noise as it moves through the natural environment!

Going for a neighborhood stroll while listening to the wind as the goal of your exploration may seem a tiny bit odd; it isn't that often that we walk along together in total silence, after all. But it can be a wonderful way to see (and hear) your local stomping grounds from a unique perspective, and it is also a pleasant way to introduce kids to the idea of taking a multisensory approach to exploring the world.

Some things to ponder as you walk and listen:

Does the wind sound different if you sit beneath a tree than it does when you are walking in the open?

What does the sound the wind is making remind you of? What words would you use to describe it to someone who isn't there with you?

Do things sound different if you cup your ear in your hand while you listen?

If you lie down in a grassy area does it change the sound the wind makes? Try the same thing under a tree.

Does the sound of the wind change as you walk around a building? What could cause it to change?

ORIGAMI
BOATS

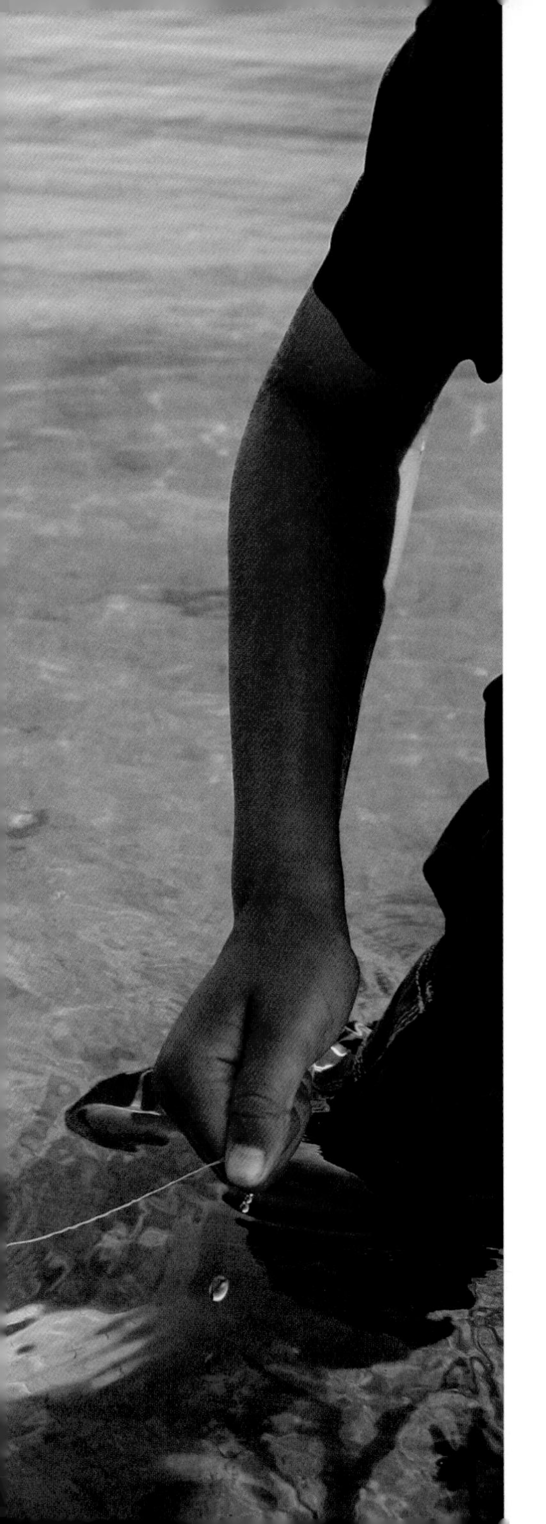

These tiny boats are based on a classic design commonly found in Japanese origami books for children. But more than just being fun to fold, these diminutive crafts make perfect vehicles for exploring the action of the wind and the way that it creates waves. Fold a fleet of these little boats, and then head to a local bay, river, creek, or stream to watch them float on the waves being made by the winds in your neighborhood.

WHAT YOU WILL NEED

- ORIGAMI PAPER
- NEEDLE AND THREAD ***(optional)***
- CANDLE AND PAINTBRUSH ***(optional)***

SCIENCE BEHIND THE SCENES: *Where Do Waves Come From?*

For many people, waves are a large part of what makes the ocean feel so captivating, active, and ever changing. Boats navigate waves, surfers ride them, and beachgoers leap over them as they crash on the shore. But where do these waves come from? What causes them? The simple answer: the wind. As the wind moves around the globe, it transfers some of its energy to the bodies of water that it crosses. When this happens, waves are formed. The size of a wave will depend on the strength of the wind and the speed at which it moves. The ocean is not the only place to be on the lookout for waves. Windy days can provide plentiful opportunities to explore waves on the shore of a lake or even moving across a large puddle.

WHAT YOU WILL DO:

Begin with a square piece of colorful origami paper. A bright color is nice for helping you to keep a close eye on your boat as it floats, even on a sunny day when there might be some glare. Fold the square in half neatly, with the right sides together. Use a folding tool to make a nice, sharp crease at the fold. [A]

Now unfold the square and then fold the outside edges of the square so that they meet the fold in the middle. Again, make a nice, sharp crease where the paper is folded. [B]

Leaving the paper folded into a rectangle, fold one of the four corners down and in so that it forms a triangle with a point that lines up with the middle of the rectangle. Repeat this process with the remaining three corners. [C]

Fold the two tips at one end of the paper in half so that they line up with the middle line and make a sharp point at the end of the paper. Repeat this process at the other end of the paper. You will find that you get a bit of a strange overlap effect in the middle of the paper, where it comes to a sort of point. Just ignore this—it's really fine. You should now have a vaguely diamond-shaped piece of paper. [D]

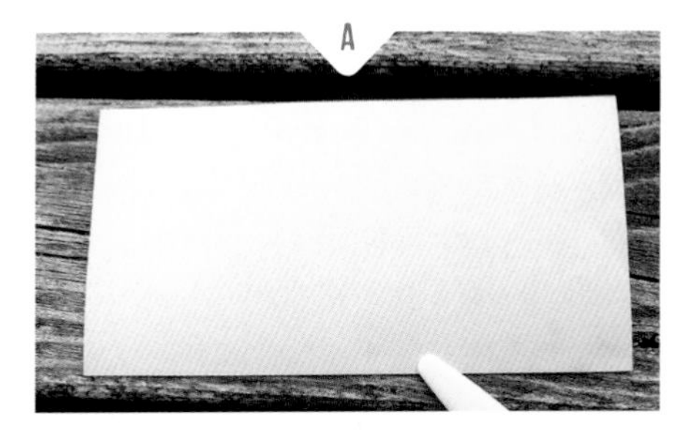

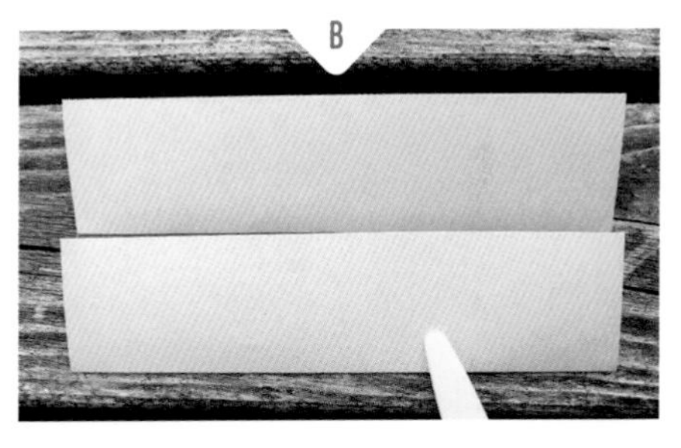

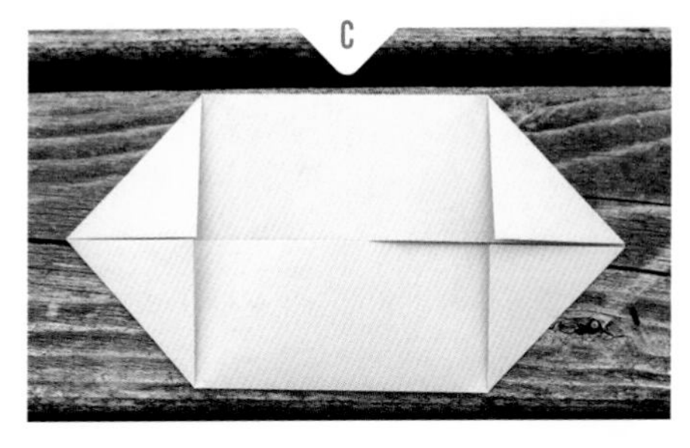

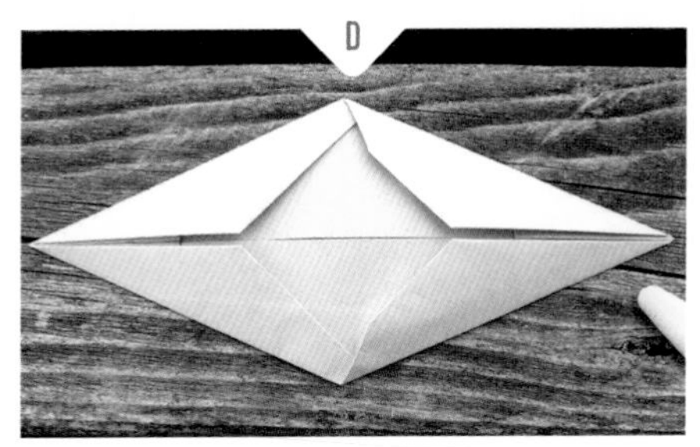

Now fold the center points of your diamond down and in, so that they line up neatly with the middle line of the paper. [E]

Gently open up the paper at the middle. You will have a nice little boat shape, but you will also notice that it needs to be turned inside out. [F]

Do this by carefully pushing the center and the sides of the boat outward with your fingers until you are able to turn the entire thing inside out. [G]

If you'd like to waterproof your boat, you can do so easily. Light a candle and wait for the center to pool with a good amount of melted wax. Using a paintbrush (be careful not to burn yourself on the flame), apply a coat of wax to the outside of the boat. This takes some patience and the wax might clump a bit on the boat, but it gives the weathered look of a boat long at sea that some mini mariners may particularly appreciate. [H]

Now all that is left to do is to take your miniature watercraft on its maiden voyage! If you are going to be outside with your boat, you might consider tying a string to one end so that it doesn't go too far downriver without you.

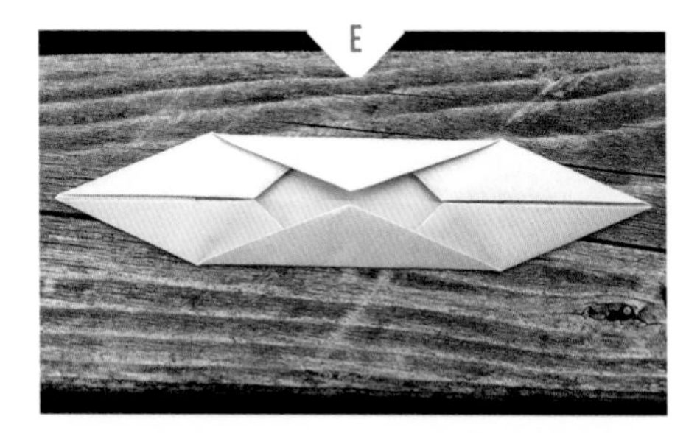

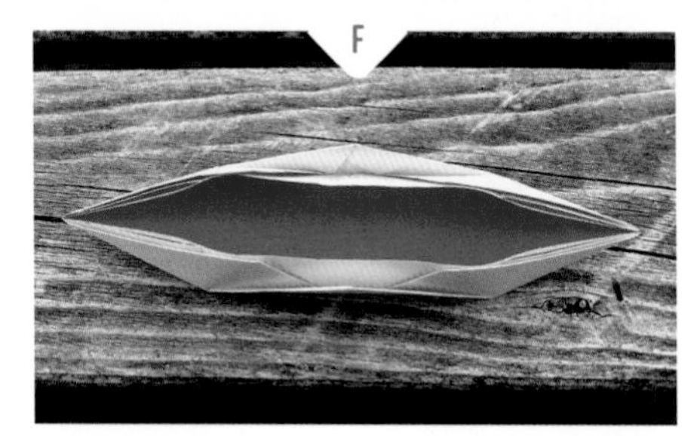

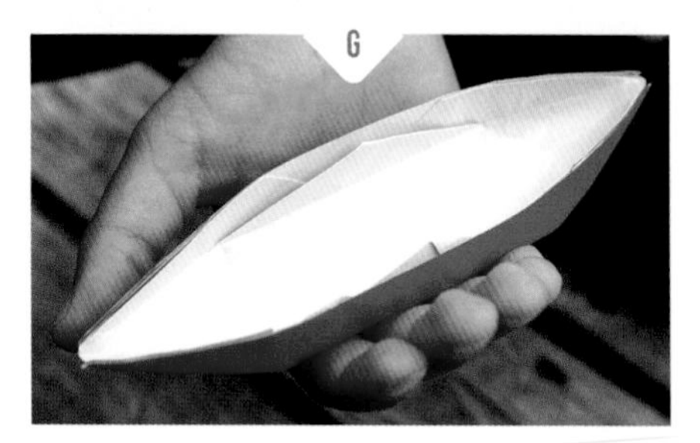

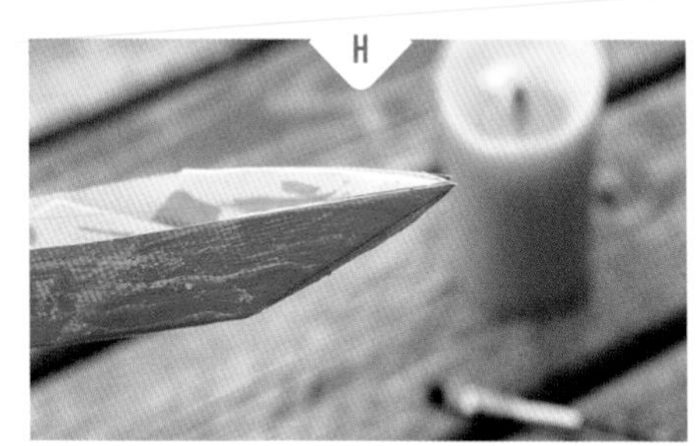

When it comes to sailing the actual seas, different-sized boats are used for different purposes and conditions. You too can make boats of different sizes by crafting boats out of paper squares of different sizes. Do the differently sized boats behave differently in the water?

Once the boat makers at your house have managed relative confidence and speed in the art of folding, make an entire fleet of boats and sail them from a common starting point at the same time. Try giving each boat a different advantage in the race (blow air on them, push them with a stick, etc.) to see how it affects their ability to sail.

SEED HEAD CLOUDS

It is inevitable for the sight of fat, fluffy clouds floating overhead to inspire a strong sense of wishing it were possible to reach out and touch them. Of course, this turns out to be difficult to do for a number of reasons. So, we offer a creative solution, with these puffy little clouds made from materials available right here on Earth. Gathering seed heads from weeds and other nearby plants that have gone to seed is a great way to explore interesting plant textures, and using those seed heads to make "clouds" in a variety of shapes and sizes allows for giving good consideration to the many forms that clouds can take.

WHAT YOU WILL NEED

- DARK-COLORED CARD STOCK OR OTHER HEAVY PAPER
- GLUE
- FLUFFY SEED HEADS LIKE DANDELION OR MILKWEED
- A ROCK TO WEIGH DOWN YOUR PAPER ON A WINDY DAY

SCIENCE BEHIND THE SCENES:
Shapes in the Clouds

From the lowest banks of thick, white fog to the largest and darkest of thunderheads rolling across the sky, clouds come in a wide variety of amazing forms.

Although there are many weather factors that conspire in cloud creation, one of the most important is the subject of the chapter at hand: the wind. The height, shape, and movement of clouds are dependent on the direction and force of the wind, as well as its speed. Generally speaking, clouds will move with the winds that prevail at the altitude where they have formed. This means that even when your feet are firmly planted on the Earth, you can make some good guesses about how windy, or not, it is in the atmosphere high, high above you.

WHAT YOU WILL DO

Begin by finding a reasonably flat surface to work on. You may want to work outside, so that you can use real clouds as a point of reference as you make your personal representations of them. You might also think about having a cloud field guide handy. There are some really fantastic ones available out there!

Once you are set up in a good work space, "draw" a cloud outline on the paper with the glue. Add more glue to the center of the cloud, so that the seeds you are using will have plenty of places to stick.

Use one of the seed heads that you have gathered to move the glue around inside the cloud shape so that the entire inside of the shape is covered with glue.

Gently tap the glue with a seed head that is full of seeds so that they begin to fall off the plant and stick to the paper. Continue to do this until the glue is covered in seeds, giving your cloud a fluffy, filled-in appearance.

The classic cumulus cloud is fun for this project, but try different types of cloud shapes as well. Talk about why clouds form in different shapes and how the wind moves clouds across the sky.

Make a collection of seed-head clouds on individual pieces of paper and once they have dried, arrange them into a book, using paper protectors to keep the seeds in place. Label the pictures to make your very own personal cloud field guide.

Press your seed-head cloud between two pieces of wax paper, sandwich the wax paper between two towels, and press with a warm iron to seal. Cut out the clouds and hang them in a window.

WHAT'S IN A (CLOUD) NAME?

Back in the early 1800s a pharmacist by the name of Luke Howard proposed the use of Latin words to classify clouds. His classifications use cloud height, shape, and possibility of precipitation to describe and name the clouds. His system is still used by both scientists and amateur weather-watchers today.

In this system, most clouds are named using two Latin words: a prefix (at the beginning), which usually tells the height of the cloud, and a suffix (at the end), which indicates the shape of the cloud.

There are ten main types of clouds that can be seen in the sky:

Cirrus	***Cumulus***
Cirrostratus	***Stratus***
Cirrocumulus	***Stratocumulus***
Altostratus	***Nimbostratus***
Altocumulus	***Cumulonimbus***

Watching the clouds float by overhead is an entertaining activity in and of itself, but if you suspect that your kids might like to take the endeavor in a more active direction, try encouraging your child to come up with a cloud classification system all their own. Perhaps special terminology for clouds that look like cats, dinosaurs, or ocean animals would be a good place to start. Or name cloud shapes after favorite friends or places you've visited as a family. You can even take photos or sketch pictures of the clouds you've observed and use them to construct a cloud field guide.

While you have your eyes turned skyward, you may see a cloud that you and your child can't quite classify. Cloud-watchers around the world continue to be on the lookout for clouds that break the rules or that don't quite fit into one of the main categories. And with more and more amateur scientists watching the skies and taking photos of the clouds above, they just might find them. In fact, the founder of the Cloud Appreciation Society, Gavin Pretor-Pinney, has put in a request with the World Meteorological Organization for the classification of a cloud he calls "undulatus asperatus," nicknamed "Jacques Cousteau clouds," for their wavelike appearance. If approved, it will be the first new type of cloud classified since 1951.

ACKNOWLEDGMENTS

Annie

Thanks first and foremost to my writing partner, personal cheerleader, and unpaid therapist, Dawn. I would never have dared to undertake this project without you by my side, and I can never thank you enough for all of your love, support, and humor, let alone your amazing ability to keep me focused.

Thank you to Erica Rand Silverman, the best agent a girl could hope for. Your ability to understand our vision and guide us in making it work on paper has been really and truly remarkable. Thank you also to Jenn Urban-Brown, whose positivity, encouragement, and excitement around this project gave us total faith in our ability to actually make it happen. All my love and gratitude to Addy, Mariam, Rowan, and Zak, for being the reason that I do all the things I do. And Ari, I am forever grateful to you for loving all four of the kids as if they were your own. Our family is better because of having you in it.

As always, I am incredibly grateful to my large and loving group of parental figures for their support. Mom, Dad, Dwight, Terri, Wendy, Jim, Craig, Lee, Robert, and Dart, thank you not only for your enthusiasm for the book but for your general enthusiasm for me.

Natalie and Annie 2.0, there is so much to say that there is almost nothing to say at all. I wouldn't have dreamed of undertaking a project like this, or much of anything else, during the last two years if I didn't know that the two of you were there to see me through it all.

And last, the most heartfelt of thanks to Jed, my partner in life and all of its various adventures. You are the reason it all works.

Dawn

I am especially grateful to have shared this process with my amazingly talented, intelligent, and exceptionally funny friend and writing partner, Annie. Thanks for keeping me in line and laughing, my friend.

Erica Rand Silverman has been a source of extraordinary support and has grounded my ideas, firmly but gently, placing them in reality.

Jennifer Urban-Brown trusted us to bring her concept for this book to life, and I cannot thank her enough for her support and guiding hand. The team at Roost Books has done an amazing job with all aspects of the project—creating the beautiful book we always imagined.

The community that has gathered together through our blog, ***Mud Puddles to Meteors***, has been wonderfully supportive, and I appreciate their continued enthusiasm for this book and sharing nature in their everyday lives.

To my tribe of friends and family, both near and far, thank you for listening, reading, and sharing countless cups of tea. My in-laws, John and Christina Smith, have shown me unwavering support, and I am forever grateful. I thank my brothers, Troy and Eric McGuirk, for the hours spent outside exploring the backyard, beach, and beyond and for their continued love (and teasing in proper doses). My parents, Gordon and Suzzanne McGuirk, have been tremendously supportive and laid the foundation for this book long ago when they sent me outside to play and allowed for hours of exploration on the shores of California. I extend my deepest gratitude for their unconditional love.

This book would have not been possible without the hard work and dedication of my husband, Wesley Smith. His love, patience, and understanding have anchored me and kept me on course.

My daughter, Fionna, has shown me the natural world with fresh eyes, and I am continually grateful to her for sharing her view of the world. My son, Dylan, has contributed to this book in ways beyond measure, and I will always be thankful he has been by my side on this journey.

RESOURCES AND FURTHER EXPLORATION

The world of weather is wide indeed, ***and as we wrote this book, we became more and more aware of just how much wild and wonderful information is available on this topic. Our hope in bringing this book to your shelves is that you and yours will use it as a jumping-off point: encouragement to delve deeper into art and science that revolve around the remarkable materials made available to us by nature. To that end, the resources below should be quite helpful. You'll find everything from field guides to picture books, making for a list of reads that are good not only for research but also for relaxing on the couch while you gather inspiration for your next foray into the outdoors, whatever the weather may be.***

Grab a Guide

Day, John A., and Vincent J. Schaefer. ***Peterson First Guide to Clouds and Weather***. 2nd ed. Boston and New York: Houghton Mifflin Harcourt, 1998.

Dunlop, Storm. ***The Weather Identification Handbook***. Guilford, CT: Globe Pequot Press, 2003.

Kahl, Jonathan D. ***National Audubon Society First Field Guide: Weather***. New York: Scholastic, 1998.

Lehr, Paul E. ***Weather (Golden Guide)***. New York: St. Martin's Press, 2001.

Mayall, R. Newton. ***The Sky Observers Guide (Golden Guide)***. New York: St. Martin's Press, 2001.

Pretor-Pinney, Gavin. ***The Cloud Collector's Handbook***. San Francisco: Chronicle Books, 2011.

Nature

Bentley, W. A. ***Snowflakes in Photographs***: Mineola, NY: Dover, 2000.

Carson, Rachel. ***The Sense of Wonder***. New York: Harper and Row, 1984.

Cherry, Lynne, and Gary Braasch. ***How We Know What We Know about Our Changing Climate: Scientists and Kids Explore Global Warming***. Nevada City, CA: Dawn Publications, 2008.

Christopher, Todd. ***The Green Hour***. Boston: Trumpeter Books, 2010.

Cornell, Joseph. ***Sharing Nature with Children***. 2nd ed., rev. and exp. Nevada City, CA: Dawn Publications, 1998. (Twentieth anniversary edition.)

Danks, Fiona, and Jo Schofield. ***Nature's Playground: Activities, Crafts, and Games to Encourage Children to Get Outdoors***. Chicago: Chicago Review Press, 2007.

Leslie, Clare Walker. ***Nature All Year Long***. New York: Greenwillow Books, 1991.

——. ***The Nature Connection: An Outdoor Workbook for Kids, Families, and Classrooms***. North Adams, MA: Storey Publishing, 2010.

Louv, Richard. ***Last Child in the Woods: Saving Our Children from Nature-Deficit Disorder***. Updated and exp. ed. Chapel Hill, NC: Algonquin Books, 2008.

——. ***The Nature Principle: Reconnecting with Life in a Virtual Age***. Chapel Hill, NC: Algonquin Books, 2012.

Obed, Ellen Bryan. ***Twelve Kinds of Ice***. Boston and New York: Houghton Mifflin Harcourt Books for Young Readers, 2012.

Russell, Helen Ross. ***Ten-Minute Field Trips***. Washington, DC: National Science Teachers Association, 1990.

Tornio, Stacy, and Ken Keffer. ***Kids' Outdoor Adventure Book: 448 Great Things to Do in Nature before You Grow Up.*** Guilford, CT: Globe Pequot Press, 2013.

——. ***We Love Nature! A Keepsake Journal for Families Who Love to Explore the Outdoors***. Boston: Roost Books, 2014.

Ward, Jennifer. ***I Love Dirt! 52 Activities to Help You and Your Kids Discover the Wonders of Nature***. Boston: Roost Books, 2008.

——. ***It's a Jungle Out There: 52 Nature Adventures for City Kids***. Boston: Roost Books, 2011.

——. ***Let's Go Outside! Outdoor Activities and Projects to Get You and Your Kids Closer to Nature***. Boston: Roost Books, 2009.

Learn More about Water and the Weather

Branley, Franklyn M. ***Down Comes the Rain***. New York: HarperCollins, 1997.

Branley, Franklyn M., and True Kelley. ***Flash, Crash, Rumble, and Roll***. New York: HarperCollins, 1999.

Cassino, Mark, and Jon Nelson. ***The Story of Snow: The Science of Winter's Wonder***. San Francisco: Chronicle Books, 2009.

Day, Trevor. ***Water***. New York: DK Publishing, 2007.

Dunlop, Storm. ***Meteorology Manual: The Practical Guide to the Weather***. Somerset, UK: Haynes Publishing, 2014.

Gibbons, Gail. ***The Reasons for the Seasons***. New York: Holiday House, 1996.

——. ***Weather Words and What They Mean***. New York: Holiday House, 1992.

Hamblyn, Richard. ***Extraordinary Clouds***. Newton Abbot, Devon, UK: David & Charles, 2009.

Locker, Thomas. ***Water Dance***. San Diego: Harcourt Brace, 1997.

Lynch, John. ***The Weather***. Buffalo, NY: Firefly Books, 2002.

Mack, Lorrie. ***Weather***. New York: DK Publishing, 2004.

McKinney, Barbara Shaw. ***A Drop around the World***. Nevada City, CA: Dawn Publications, 1998.

Patkau, Karen. ***Who Needs an Iceberg?*** New York: Tundra Books, 2012.

Strauss, Rochelle. ***One Well: The Story of Water on Earth***. Toronto: Kids Can Press, 2007.

Wick, Walter. ***A Drop of Water: A Book of Science and Wonder***. New York: Scholastic, 1997.

Arts and Crafts

Davies, Jacqueline. ***The Boy Who Drew Birds: A Story of James John Audubon***. Boston: Houghton Mifflin, 2004.

Dickens, Rosie. ***The Usborne Art Treasury: Pictures, Paintings, and Projects***. London: Usborne, 2006.

Leslie, Clare Walker, and Charles E. Roth. ***Keeping a Nature Journal***. North Adams, MA: Storey Publishing, 2000.

Martin, Laura. ***Nature's Art Box: From T-Shirts to Twig Baskets, 65 Cool Projects for Crafty Kids to Make with Natural Materials You Can Find Anywhere***. North Adams, MA: Storey Publishing, 2003.

Van't Hul, Jean. ***The Artful Parent: Simple Ways to Fill Your Family's Life with Art and Creativity***. Boston: Roost Books, 2013.

Tell Me a Story

COLD DAYS

Baker, Keith. ***No Two Alike***. New York: Beach Lane Books, 2011.

Clark, Joan. ***Snow***. Berkeley, CA: Groundwood Books, 2006.

Denslow, Sharon Phillips. ***In the Snow***. New York: Greenwillow Books, 2005.

Lee, Wong Herbert. ***Tracks in the Snow***. New York: Square Fish, 2007.

Martin, Jacqueline Briggs. ***Snowflake Bentley***. Boston: Houghton Mifflin, 1998.

Stewart, Melissa, and Constance R. Bergum. ***Under the Snow.*** Atlanta: Peachtree Publishers, 2009.

RAINY DAYS

Bauer, Caroline Feller, ed. ***Rainy Day: Stories and Poems***. Philadelphia: Lippincott Williams & Wilkins, 1986.

Hesse, Karen, and Jon J Muth. ***Come on, Rain!*** New York: Scholastic, 1999.
Serfozo, Mary. ***Rain Talk***. New York: Aladdin Books, 1993.

SUNNY DAYS

Dotlich, Rebecca Kai, and Jan Gilchrist. ***Lemonade Sun: And Other Summer Poems***. Honesdale, PA: Wordsong, 2001.
Perkins, Lynne Rae. ***Pictures from Our Summer Vacation***. New York: Greenwillow Books, 2007.
Ringgold, Faith. ***Tar Beach***. New York: Dragonfly Books, 1996.
Rocco, John. ***Blackout.*** New York: Disney-Hyperion, 2011.
Rylant, Cynthia, and Stephen Gammell. ***The Relatives Came.*** Columbus, Ohio: Modern Curriculum Press, 1993.

WINDY DAYS

Berger, Carin. ***The Little Yellow Leaf.*** New York: Greenwillow Books, 2008.
Cobb, Vicki, and Julia Gorton. ***I Face the Wind***. New York: HarperCollins, 2003.
dePaola, Tomie. ***The Cloud Book***. Reprint ed. New York: Holiday House, 1984.
Dorros, Arthur. ***Feel the Wind***. Reprint ed. New York: HarperCollins, 2000.
Stoop, Naoko. ***Red Knit Cap Girl to the Rescue***. New York: Little, Brown, 2013.
Wiesner, David. ***Sector 7***. New York: Clarion Books, 1999.

GENERAL NATURE

Berne, Jennifer. ***On a Beam of Light: A Story of Albert Einstein***. San Francisco: Chronicle Books, 2013.
Cole, Henry. ***On Meadowview Street***. New York: Greenwillow Books, 2007.
Locker, Thomas. ***Sky Tree: Seeing Science through Art***. New York: HarperCollins, 1995.

McClure, Nikki. ***Mama, Is It Summer Yet?*** New York: Abrams Books for Young Readers, 2010.

Pearson, Susan. ***Silver Morning***. San Diego: Harcourt Brace, 1998.

Stead, Philip C. ***Bear Has a Story to Tell***. New York: Roaring Brook Press, 2012.

Yolen, Jane. ***Owl Moon***: New York: Philomel Books, 1987.

Tools for Young Naturalists

The Old Farmer's Almanac Weather Watcher's Calendar. Dublin, NH: Yankee Publishing.

BioQuipBugs: **www.bioquipbugs.com**

Imagine Childhood: **http://store.imaginechildhood.com**

Acorn Naturalists: **www.acornnaturalists.com**

Supplies for Young Artists

Dick Blick: **www.dickblick.com**

Nova Natural Toys and Crafts: **www.novanatural.com**

Other Resources on the Web

American Museum of Natural History. "Water: H2O = Life." **www.amnh.org/exhibitions/water**

The Artful Parent. **http://artfulparent.com**

Children and Nature Network. **www.childrenandnature.org**

Cloud Appreciation Society. **http://cloudappreciationsociety.org**

NASA. "For Kids Only: Earth Science." **http://kids.earth.nasa.gov**

National Wildlife Federation. "Be Out There." **www.nwf.org/Be-Out-There.aspx**

The Nature Conservancy. "Nature Rocks." **www.naturerocks.org**

SnowCrystals.com. **www.its.caltech.edu/~atomic/snowcrystals/**

ABOUT THE AUTHORS

Annie Riechmann is an educator and unabashed nature lover. Her many years of experience in education have taken her from coast to coast, where she has taught in both rural and urban settings. These experiences have given her a unique perspective on ways that families can connect to nature, no matter where they live. Annie is the creator of ***Alphabet Glue,*** a literacy-based e-magazine for families, and she is an advocate for outdoor education in the public schools. She is the mother of young naturalists herself and is not shy about her love of kitchen table science! She lives in Massachusetts with her family.

Dawn Suzette Smith is a self-taught naturalist and trained educator. For the past fifteen years she has worked to promote children's connection with nature and lifelong love of outdoor pursuits for both physical and mental health. Her writing and photography have been featured in various print and online magazines and books, and in exhibits with the National Park Service. Dawn currently homeschools two curious nature lovers and leads nature walks year-round to help families connect with nature through child-led nature study in the wild forests and along the rugged coastline of Nova Scotia, Canada. She can often be found under a shade tree with her kids watching clouds pass by and shadows dance.

Annie and Dawn blog together at ***Mud Puddles to Meteors.***